Lecture Notes in Computer Science 778

Edited by G. Goos and J. Hartmanis

Advisory Board: W. Brauer   D. Gries   J. Stoer

M. Bonuccelli   P. Crescenzi
R. Petreschi (Eds.)

Algorithms and Complexity

Second Italian Conference, CIAC '94
Rome, Italy, February 23-25, 1994
Proceedings

Springer-Verlag

Berlin Heidelberg New York
London Paris Tokyo
Hong Kong Barcelona
Budapest

Series Editors

Gerhard Goos
Universität Karlsruhe
Postfach 69 80
Vincenz-Priessnitz-Straße 1
D-76131 Karlsruhe, Germany

Juris Hartmanis
Cornell University
Department of Computer Science
4130 Upson Hall
Ithaca, NY 14853, USA

Volume Editors

Maurizio Bonuccelli
Pierluigi Crescenzi
Rossella Petreschi
Dipartimento di Scienze dell'Informazione
Università degli Studi di Roma "La Sapienza"
Via Salaria 113, I-00198 Rome, Italy

CR Subject Classification (1991): F.1-2, E.1

ISBN 3-540-57811-0 Springer-Verlag Berlin Heidelberg New York
ISBN 0-387-57811-0 Springer-Verlag New York Berlin Heidelberg

CIP data applied for

© Springer-Verlag Berlin Heidelberg 1994
Printed in Germany

Typesetting: Camera-ready by author
SPIN: 10131934   45/3140-543210 - Printed on acid-free paper

Preface

The papers in this volume were presented at the Second Italian Conference on Algorithms and Complexity (CIAC '94). The conference took place February 23-25, 1994, in Rome, Italy and it is intended to be a biannual series of international conferences that present research contributions in theory and applications of sequential, parallel, and distributed algorithms, data structures, and computational complexity.

In response to the organizing committee's call for papers, 32 papers were submitted. From these submissions, the program committee selected 14 for presentation at the conference. Each paper was evaluated by at least four program committee members. In addition to the selected papers, the organizing committee invited the following people to give plenary lectures at the conference: Juris Hartmanis, Silvio Micali, Roberto Tamassia, Uzi Vishkin, and Mihalis Yannakakis.

We wish to thank all members of the program committee, the other members of the organizing committee, the five plenary lecturers who accepted our invitation to speak, all those who submitted abstracts for consideration, and all people who reviewed papers at the request of program committee members.

Rome, February 1994

M. Bonuccelli
P. Crescenzi
R. Petreschi

Conference chair:

M. Bonuccelli (Rome)

Program Committee:

G. Ausiello (Rome)
A. Bertoni (Milan)
G. Bilardi (Padua)
D. Bovet (Rome)
S. Even (Haifa)

J. van Leeuwen (Utrecht)
F. Luccio (Pisa)
C. Papadimitriou (San Diego)
M. Minoux (Paris)
P. Spirakis (Patras)

Additional Referees:

R. Bar-Yehuda
A. Bertossi
E. Biham
D. Bruschi
A. Campadelli
N. Cesa-Bianchi
E. Dinitz
T. Etzion
P. Ferragina
P. Franciosa
E. Feuerstein
F. Gadducci
L. Gargano
J. Glasgow

M. Goldwurm
S. Katz
E. Kushilevitz
A. Itai
S. Leonardi
A. Litman
J. Makowsky
A. Mancini
A. Marchetti-Spaccamela
L. Margara
G. Mauri
C. Mereghetti
S. Moran
L. Pagli

S. Pallottino
E. Petrank
G. Pighizzini
Y. Pnueli
R. Posenato
G. Pucci
S. Rajsbaum
F. Ravasio
A. Roncato
H. Shachnai
A. Shirazi
M. Tarsi
M. Torelli

Sponsored by:

Dipartimento di Scienze dell'Informazione, Università di Roma "La Sapienza"
Università degli Studi Roma "La Sapienza"
CNR – Comitato Scienze Matematiche
CNR – Comitato Scienze e Tecnologia dell'Informazione
Bagnetti s.r.l.
Cassa Rurale ed Artigiana di Roma
Credito Artigiana
ENEA
Sun Microsystems Italia S.P.A

Table of Contents

Invited Presentations

On the Intellectual Terrain Around NP   1
J. Hartmanis, S. Chari

Advances in Graph Drawing   12
A. Garg, R. Tamassia

On a Parallel-Algorithms Method for String Matching Problems   22
S.C. Sahinalp, U. Vishkin

Some Open Problems in Approximation   33
M. Yannakakis

Regular Presentations

New Local Search Approximation Techniques for Maximum Generalized
Satisfiability Problems   40
P. Alimonti

Learning Behaviors of Automata from Multiplicity and Equivalence Queries   54
F. Bergadano, S. Varricchio

Measures of Boolean Function Complexity Based on Harmonic Analysis   63
A. Bernasconi, B. Codenotti

Graph Theory and Interactive Protocols for Reachability Problems on
Finite Cellular Automata   73
A. Clementi, R. Impagliazzo

Parallel Pruning Decomposition (PDS) and Biconnected Components of Graphs   91
E. Dekel, J. Hu

A Non-Interactive Electronic Cash System   109
G. Di Crescenzo

A Unified Scheme for Routing in Expander Based Networks   125
S. Even, A. Litman

Dynamization of Backtrack-Free Search for the Constraint Satisfaction Problem   136
D. Frigioni, A. Marchetti-Spaccamela, U. Nanni

Efficient Reorganization of Binary Search Trees 152
M. Hofri, H. Shachnai

Time-Message Trade-Offs for the Weak Unison Problem 167
A. Israeli, E. Kranakis, D. Krizanc, N. Santoro

On Set Equality-Testing 179
T.W. Lam, K.H. Lee

On the Complexity of Some Reachability Problems 192
A. Monti, A. Roncato

On Self-Reducible Sets of Low Information Content 203
M. Mundhenk

Lower Bounds for Merging on the Hypercube 213
C. Rüb

On the Intellectual Terrain around NP[†]

Juris Hartmanis[1] and Suresh Chari[2]

[1] Cornell University and Max-Planck Institut für Informatik
[2] Cornell University

Abstract. In this paper we view $P \overset{?}{=} NP$ as the problem which symbolizes the attempt to understand what is and is not feasibly computable. The paper shortly reviews the history of the developments from Gödel's 1956 letter asking for the computational complexity of finding proofs of theorems, through computational complexity, the exploration of complete problems for NP and PSPACE, through the results of structural complexity to the recent insights about interactive proofs.

1 GÖDEL'S QUESTION

The development of recursive function theory, following Gödel's famous 1931 paper [Gö31] on the incompleteness of formal mathematical systems, clarified what is and is not effectively computable. We now have a deep understanding of effective procedures and their absolute limitations as well as a good picture of the structure and classification of effectively undecidable problems. It is not clear that the the concept of feasible computability can be defined as robustly and intuitively satisfying form as that of effective computability. At the same time it is very clear that with full awareness of the ever growing computing power we know that there are problems which remain and will remain, in their full generality, beyond the scope of our computing capabilities. One of the central tasks of theoretical computer science is to contribute to a deeper understanding of what makes problems hard to compute, classify problems by their computational complexity and yield a better understanding of what is and is not feasibly computable.

It is interesting to observe that the effort to understand what is and is not effectively computable was inspired by questions about the power of formal systems and questions about theorem proving. Not too surprisingly, the questions about the limits of feasible computability are also closely connected to questions about the computational complexity of theorem proving. It is far more surprising and interesting that it was again Gödel, who's work had necessitated the investigations of what is effectively computable, who asked the key question about feasible computability in terms of theorem proving. In a most interesting 1956 letter to his colleague at the Institute for Advanced Study, John von Neumann, Gödel asks for the computational complexity of finding proofs for theorems in formal systems(cf. [Har89]). Gödel is very precise, he specifies the Turing machine as a computational model and then asks for the function which bounds the number of steps needed to compute proofs

---

[†] This work is supported in part by NSF grant CCR–9123730

of length n. It does not take very long to realize that Gödel was, in modern terminology, asking von Neumann about the deterministic computational complexity of theorem proving and thus about the complexity of NP. In the same letter Gödel also asks about the computational complexity of problems such as primality testing and quite surprisingly, expresses the opinion that the problem of theorem proving may not be so hard. He mentions that it is not unusual that in combinatorial problems the complexity of the computation can be reduced from the N steps required in the brute force method to $\log N$ steps, or linear in the length of the input. He mentions that it would not be unreasonable to expect that theorem proving could be done in a linear or quadratic number of steps. Strange that the man who showed the unexpected limits of formal theorem proving did not seem to suspect that computational theorem proving and therefor NP may be at the limits of feasible computability. It is very unfortunate that von Neumann was already suffering from cancer and passed away the following year. No reply to this letter has been found and it seems that Gödel did not pursue this problem nor try to publicize it.

2 Complexity theory and complete problems

The full understanding of the importance of Gödel's question had to wait for the development of computational complexity which was initiated as an active computer science research area in the early 1960's. The basic hierarchy results were developed, the key concept of the complexity class, the set of all problems solvable(or languages acceptable) in a given resource bound, was defined and investigated[HS65, HLS65]. In a very general sense, the most important effect of this early work in complexity theory was the demonstration that there are strict quantitative laws that govern the behavior of information and computation. A turning point in this work came in 1971 with Cook's proof that SAT, the set of satisfiable Boolean formulas, was complete for NP, the class of nondeterministic polynomial time computations [Coo71](these results were independently discovered a year later by Levin [Lev73]). In other words, any problem that can be solved in polynomial time by guessing a polynomially long solution and then verifying that the correct solution had been guessed, could be reduced in polynomial time to SAT. In 1972 Karp [Kar72] showed that many combinatorial problems were in NP and could be reduce to SAT. On of the key problems in this set was the Clique decision problem: given a graph G and an integer k, determine if there is a clique of size k in this graph. Cook's and Karp's work virtually opened a flood gate of complete problems in NP and PSPACE. From all areas of computer science mathematics, operations research etc., came complete problems in all sizes and shapes all reducing to SAT and SAT reducing to them (see [GJ79] for a compendium).

Later, many other complexity classes were defined and found to have complete languages or problems. Among the better known ones are SPACE($\log n$), NSPACE($\log n$), P, NP, SPACE(n), NSPACE(n), the Polynomial Hierarchy(PH), PSPACE, NPSPACE(which is the same as PSPACE), EXPTIME, NEXPTIME, and EXPSPACE. The reassuring aspect of these classes is that they are very robust in their definition and that they appear naturally in many other settings which may or may not be directly connected to computing. For example, a variety of these complexity classes appear naturally as the classes definable by different logics, without any direct reference to computing(cf. [Imm89] for a survey). There is no doubt that complexity classes are a key concept and that they reveal a natural and important structure in the classification of computational problems and mathematical objects.

3 Structural Complexity Theory

The exploration of the structure of the complexity of computations has been an active area of research for more than a decade. This research explores the relations between various complexity classes and it investigates the internal structure of individual complexity classes. Figure 1 presents the key complexity classes below EXPSPACE, the class of all problems solvable with tape size bounded by an exponential in the length of the input. Besides the space and time bounded classes such as P, PSPACE etc. mentioned earlier the figure also shows the second level of the Polynomial Hierarchy in detail. $P^{SAT||[k]}$ is the class of languages recognizable by polynomial time oracle machines which can make up to k non-adaptive queries to a SAT oracle. The language of uniquely satisfiable formulas, USAT, is in Σ_2^p, the second level of the PH and also the complement of USAT is in Σ_2^p. In fact, USAT, is in the class $P^{SAT||[2]}$. $P^{SAT[\log n]}$ is the class of languages accepted by polynomial time oracle machines which make logarithmic number of queries to SAT and P^{SAT} is the class of languages recognizable with a polynomial number of queries to SAT. For several NP optimization problems where the optimum value is bounded by a polynomial in the input size, such as the Clique problem (the maximum clique size is the number of vertices), the optimum value can be computed with logarithmically many queries to SAT, by a binary search on the polynomial sized range of possible values. Optimization problems such as the Traveling Salesperson Problem can be solved with polynomially many queries to SAT.

Though we know a tremendous amount about the relations between these complexity classes, we still have no proof that they all are distinct: There is no proof that P is not equal to PSPACE! There is a very strong conviction that all the major complexity classes are indeed different, but in spite of all our efforts there is no nontrivial separation result. A major result would be any separation of: P and NP, NP and PSPACE, the various levels of the Polynomial Hierarchy, PSPSACE and EXPTIME, EXPTIME and NEXPTIME, and the last two from EXPSPACE. Less dramatic, but still very interesting would be the separation of the lower complexity classes, including SPACE($\log n$) and P or some even lower classes (defined by circuit models) not discussed here. It is clear that there are major problems facing complexity theory on which no substantial progress has been made since their definition, in some cases more than twenty years ago. The successes in recursive function theory have come from diagonalization arguments whose sharpness and sophistication has been elevated to a fine art. Unfortunately, though structural complexity theory has been strongly influenced by concepts from recursive function theory, diagonalization proof techniques have failed to dent the notorious separation problems. Since we can not diagonalize over these classes to construct a language just outside of the class, *i.e.* in a higher but not too high a class, we lack proof techniques to be able to deal with all possible ways of computing a problem in a given class to show that a certain problem from the class above can not be so computed. We badly need new concepts

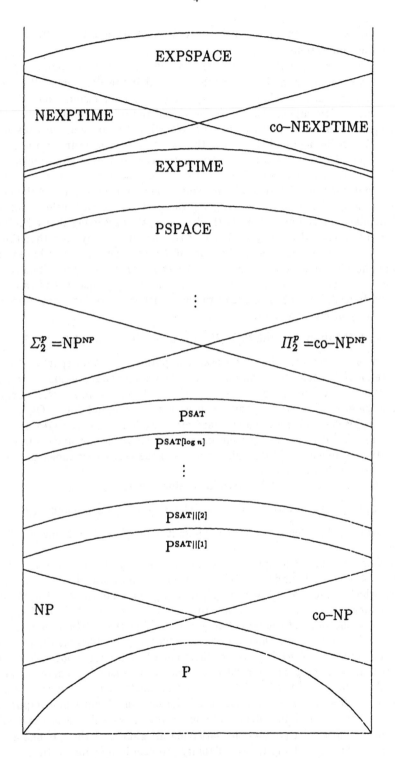

Fig. 1. Key complexity classes below EXPSPACE

and techniques to reason about all the possible computations in a complexity class to show their limitations.

Given the current situation, a working hypothesis in structural complexity theory is that P is different than NP and further more that the Polynomial Hierarchy is infinite. In particular, there have been many very interesting results which show that a given assumption implies that PH is finite, and thus giving in our view, a strong indication that the assumption is not true. We will illustrate some such results.

In 1977 Len Berman and the first author were led to conjecture on the strength of many special cases and with analogy to the corresponding situation in recursive function theory, that all NP complete sets are isomorphic under polynomial time computable isomorphisms [BH77]. This means that for any two NP complete problems there exists a bijection, reducing the problems to each other, and which is polynomial time computable in both directions, thus asserting that all NP complete problems are indeed very similar in structure. It is easily seen that this, the Berman–Hartmanis conjecture, implies that all NP complete sets must have roughly the same density. If we refer to a set as sparse if it has only polynomially many elements up to size n, then the conjecture implies that no sparse sets can be NP complete. Indeed, a former student of the first author, Steve Mahaney, proved the following very interesting result.

Theorem 1 (Mahaney [Mah82]). *If a sparse set is* NP *complete then* P=NP.

This leads us to accept, using our working hypothesis, that no sparse set can be complete for NP. Thus, there can not be a polynomially large dictionary which we can consult to solve an NP complete problem in polynomial time with its help. We describe the gist of the proof of this result with a new method due to Ogiwara and Watanabe [OW90], which has made a hard proof [Mah82] quite manageable. As you will see, the key ideas are deceptively simple, but please do not underestimate the originality and difficulty of the original proof and the very clever choice of the right NP complete set in the new proof.

Proof Outline: Instead of SAT, we will consider the set

$$A = \{(x, F) \mid F \text{ an } r-\text{variable formula}, |x| = r, \text{ and } (\exists y) \; x < y \text{ and } F(y) = 1\}.$$

It is easy to see that this set is NP complete; it is in NP and by fixing $x = 00....0$, it is SAT.

Let S be a sparse NP complete set, and let the density function of S, i.e. $|\{x|x \in S \text{ and } |x| \leq n\}|$, be bounded by n^k. Let f be the polynomial time computable function which reduces A to S with $|f(x)| \leq N(|x|)$. Think of the 2^r r-length binary strings arranged on a line in increasing order. Divide this line in $2(N(n))^k$ segments and compute the value of the reduction f of A to S at the $2(N(n))^k$ points at the beginning of each segment. Should all these values of f be different then we know that among the first $(N(n))^k + 1$ values there must be one that is not in S and thus shows that there is no y to the right of it which can satisfy F. Therefore we can remove the right half of the line since no solution exists there. We now repeat this process on the first half. Clearly, if our luck holds and, in all following computations the f values are distinct, in polynomially many rounds we will reach polynomially many values which must contain the satisfying assignment if there is one and we check all possibilities, solving the satisfiability problem in polynomial time.

Clearly, we can not expect that all the f values will be distinct all the time. At the same time, equal values also contain a lot of good information and can be used to eliminate from consideration the segments between any pair of values with same function value. To see this, observe that if for $x < x'$, we have $f(x) = f(x')$ not in S, then there is no solution to the right of x and therefore no solution between x and x'. If $f(x) = f(x')$ is in S, there are solutions to the right of x', and hence we can eliminate the segment between x and x' without losing all solutions. Again it is seen that if there are many equal values we toss out all the segments between pairs of equal values and the line is shortened. It is not too hard to see that combining both methods the total length of the remaining segments to be searched decreases by at least half each time and we can determine in polynomial time if a the formula F is satisfiable, which implies that P=NP.

To illustrate an assumption which forces the Polynomial Hierarchy to be finite we will consider deterministic polynomial time computations which can query a SAT-oracle, *i.e.* during the computation a polynomial number of questions can be asked about the outcome of NP computations. These considerations will also lead us to some recent interesting results about probabilistic computations which are intuitively very satisfying. To recall, $P^{\text{SAT}|||[k]}$ denotes the class of polynomial time computations with k parallel *i.e.* non-adaptive queries to the SAT oracle.

Theorem 2 (Kadin [Kad88]). *If for any k,* $P^{\text{SAT}|||[k-1]} = P^{\text{SAT}|||[k]}$*, then the Polynomial Hierarchy is finite.*

Not only can sparse sets not be NP complete if P is not equal to NP, but if the PH is infinite as we strongly expect, there is no way of even saving a single query to SAT out of millions of queries. On the other hand, our intuition suggests that as the number of queries to SAT increases the value of additional queries should decrease. Kadin's result shows that this is not the case, but there is another way of looking at this problem which indeed justifies our intuition of the decreasing value of queries as the numbers increases. To see this we consider randomized reductions:

Definition 3. A language $H \leq_m^{rp}$ reduces to language K, with probability p if there exists a polynomial time function f such that

$$x \in H \Rightarrow \text{Prob}[f(x,z) \in K] \geq p$$

$$x \notin H \Rightarrow \text{Prob}[f(x,z) \notin K] = 1$$

when z is chosen at random from $q(n)$ long binary strings, where q is a polynomial.

The $\leq_m^{bpp}$ (two-sided error) reduction is defined similarly, with the condition being

$$\text{Prob}[x \in H \text{ iff } f(x,z) \in K] > p.$$

The question now is, with what probability can languages in $P^{\text{SAT}|||[k]}$ be reduced to $P^{\text{SAT}|||[k-1]}$. A very recent result by a former student of the first author, Pankaj Rohatgi, shows that there are very interesting threshold results about randomized reductions.

Theorem 4 (**Rohatgi [Roh92]**). *Every language in* $\mathrm{P}^{\mathrm{SAT}||[k]}$ *can be reduced to a language in* $\mathrm{P}^{\mathrm{SAT}||[k-1]}$ *with two-sided error reductions of probability* $1 - \frac{1}{k+1}$. *And this result is optimal: if there is a such a reduction possible with probability* $1 - \frac{1}{k+1} + \frac{1}{poly(n)}$ *then the* PH *is finite, where n is the size of the input to the reduction.*

This is a very interesting result: First it shows that with increasing k the value of additional queries to SAT decreases since the probability that it reduces to a problem solvable by one less query increases as k grows. At a million queries to SAT an additional query indeed has a small value. It also shows, surprisingly, that in randomized reductions there are very sharply defined threshold effects such that if it is possible to pass this threshold then the polynomial hierarchy collapses, just as in the deterministic case, according to Kadin's result. It is also interesting to note that the proof of this result uses the 'hard-easy' method used in Kadin's proof in a new setting with technically interesting methods. The results for the one-sided error case are similar but because of the stricter reductions the threshold probability is $1 - \frac{1}{\lceil k/2 \rceil}$. Again the violation of this threshold forces PH to be finite. The delightful part in the proofs is that it the reductions that achieve the desired probability bound are simple and it is surprising that the 'natural' method is also an optimal one. We urge you to read Rohatgi's PhD dissertation[Roh94].

4 Interactive Proofs

In 1985 two models of interactive randomized computation were introduced with very different motivations. In order to extend NP class slightly to capture some problems not known to be in NP, Babai defined the Arthur–Merlin games[Bab85](see also [BM88]). Motivated by possible applications to cryptographic protocols, Goldwasser, Micali and Rackoff defined the class of languages with interactive proofs[GMR89]. To motivate the introduction to this computing model, let us recall that NP is simply that class of problems for which, when a solution is suggested, it can be verified in polynomial time that the solution indeed solves the problem. We can think of this as a very powerful prover seeing the problem posed to the verifier and then simply sends him the solution, which than can be checked for validity by the verifier. Also, if no such solution exists, no prover, however powerful, can convince the verifier of the existence of a solution. Here the communication is one-way from the prover to the verifier. It is easily seen that adding a two way interaction in this model does not increase the computing power. To see this, we just have to observe that the verifier is a deterministic polynomial time machine and that the very powerful prover can compute the questions which the verifier will ask upon seeing the posed problem and in response to the answers from the old prover. Thus the prover after seeing the problem can send the whole exchange of questions and answers to convince the verifier that there is indeed a solution to the posed problem, and we are back to NP case. To have a unpredictable interaction between the prover and the verifier we need to add randomness to the arsenal of the verifier and give up the total certainty of verification of the solution as in the case of NP case. This leads us to the following definition of languages with interactive proofs.

Definition 5 ([GMR89]). A language L is in IP (*i.e.* it has an interactive proof) if there exists a verifier V, a randomized polynomial time machine such that

$$x \in L \;\Rightarrow\; \text{There exists a prover } P^* \text{ such that } \text{Prob}[(V, P^*) \text{ accept on } x] = 1$$

$$x \notin L \;\Rightarrow\; \text{For all provers } P \text{ Prob}[(V, P) \text{ accept on } x] < \frac{1}{3}.$$

Note that repeated, independent executions of this protocol reduces the probability of failure exponentially fast. For some time after the introduction of these concepts it was not clear how powerful interactive proofs were. There is a very nice proof that the graph non-isomorphism problem, which is in co-NP and is not known to to be in NP, has an interactive proof [GMW86]. The protocol is indeed simple: For any pair of graphs (H, K), the Verifier selects randomly H or K, randomly permutes the graph, and asks the Prover to identify the graph as H or K. If H and K are not isomorphic a honest prover has no difficulty identifying the graph. If they are isomorphic, no Prover is capable of distinguishing between the randomly permuted versions of H and K, which are isomorphic. Thus, no prover can do any better than guessing the answer and hence cannot deceive the verifier with probability more than $\frac{1}{2}$. Again, by a repetition of this question-answer exchange, the probability that the verifier is fooled by a dishonest prover can be made exceedingly small.

It is also interesting to recall that there were oracle results[FS88] which gave relativized worlds in which co-NP was not contained in IP. This suggested, incorrectly, that IP was not very powerful or, according to the heuristic use of oracle results, indicating that it should be very hard to prove that co-NP is contained in IP in the real (unrelativized) world. The oracle hueristic implied that such a proof can not be obtained with "known methods".

The determination of the full power of IP came very suddenly and it was a great surprise. In 1990, Lund, Fortnow, Karloff and Nisan [LFKN90] developed the techniques to show that the entire polynomial hierarchy is contained in IP and Shamir [Sha90] finished the characterization of the power of interaction and randomization by showing that IP=PSPACE. This was indeed a startling surprise with further unexpected consequences. The proof exploits in a very elegant way our better understanding of polynomials than Boolean formulas. The fact that polynomials can be 'fingerprinted' be evaluating them at a random point, was the key to the very ingenious proof. We know that if two polynomials, p and q of low degree evaluate to the same value with high probability at a randomly chosen point x, then, with high probability they are identical polynomials, since x is a root of the low degree polynomial $p(x) - q(x)$, and such polynomials have few roots unless identically zero. The key idea in the proof is to replace a given quantified Boolean formula, which constitute a PSPACE complete problem, QBF, by operations over a field of integers modulo a large prime. To convert a QBF into a polynomial, 'or' is replaced by $+$, 'and' by $\times$, 'not'x by $(1-x)$, 'for all'x by $\prod_{x=0,1}$, and 'there exists'x by $\sum_{x=0,1}$. After that, a clever sequence of questions about the resulting polynomials and their evaluation (fingerprinting) at random points, yields the interactive protocol to test if the given QBF is satisfiable, thus showing that QBF$\in$IP, and since QBF is complete for PSPACE, PSPACE$\subseteq$IP. This startling result also gives a natural counterexample to refute the random oracle hypothesis. It is shown in [HCRR90b] that with probability 1, for a random oracle A, $IP^A \neq PSPACE^A$. The IP=PSPACE result also allows one to tie the *width* of a proof of a theorem in a formal system to the ease of convincing a verifier of its correctness with high probability[HCRR90a]. Informally, we can think of a proof of a theorem as being written on a two–dimensional page so that it can be verified easily, for *e.g.* by a finite automaton which can read one symbol on 2 adjacent lines of the proof. This can be shown to be a robust definition with the finite automaton replaced by several other models, all yielding the same class of proofs. It can then be shown that PSPACE is the class of languages with proofs with polynomially wide proofs, where the width of a 2–dimensional proof being the largest number of symbols on any one line. IP=PSPACE thus implies that width, as opposed to the length, of the proof determines how quickly one can give overwhelming evidence that a theorem is provable without showing its full proof

The impact of the IP=PSPACE result is still being explored in complexity theory but some related results followed in rapid succession. Motivated again by their possible application in cryptographic protocols, Ben-Or, Goldwasser, Kilian and Wigderson [BGKW88] had earlier proposed, MIP, a multi–prover version of interactive proofs. Babai, Fortnow and Lund [BFL90] showed that this model, which has two provers who do not communicate with each other, increases the computational power substantially: MIP = NEXPTIME. Fortnow, Rompel and Sipser [FRS88] showed that the MIP model was equivalent to the following oracle model

Definition 6. A language L is in MIP if there exists a polynomial time probabilistic oracle machine $V($ the verifier $)$ such that

$$x \in L \;\Rightarrow\; \text{There exists oracle } O \text{ such that Prob}[V^O(x) \text{ accepts }] = 1$$

$$x \in L \;\Rightarrow\; \text{For all } O \text{ Prob}[V^O(x) \text{ accepts }] \leq \frac{1}{3}$$

Since the oracle answers can be thought of as an exponentially long 'proof' of membership, the result MIP=NEXPTIME, says that for languages in exponential time there are exponentially long proofs of membership which can be checked by inspection at a polynomial number of places. This characterization of MIP and the MIP=NEXPTIME were surprisingly used to show the hardness of approximating the clique problem[FGL+91] unless EXPTIME=NEXPTIME. This also initiated the attempt to 'scale down' the result on proofs whose correctness can be checked by inspection at very few places[BFLS91]. Two big breakthroughs followed: The results of [AS92] showed that indeed languages in NP had proofs which could be verified by checking at logarithmically many (in fact even smaller) places of the proof. Finally Arora *et al.* [ALM+92] showed that NP is the set of languages which have proofs which can be checked by verifiers which use logarithmically many random bits and which inspect the proof at only a constant number of positions! This surprising characterization has led to many results on the hardness of approximating many combinatorial problems[ALM+92, LY93, ABSS93], whose approximability has been open for several years.

References

[ABSS93]   S. Arora, L. Babai, J. Stern, and Z. Sweedyk. The hardness of approximate optima in lattices, codes and systems of linear equations. In *Proceedings of the 34th IEEE Symposium on Foundations of Computer Science*, pages 724–733, 1993.

[ALM$^+$92]   S. Arora, C. Lund, R. Motwani, M. Sudan, and M. Szegedy. Proof verification and the hardness of approximation problems. In *Proceedings of the 33rd IEEE Symposium on Foundations of Computer Science*, pages 14–23, 1992.

[AS92]   S. Arora and S. Safra. Probabilistic checking of proofs. In *Proceedings of the 33rd IEEE Symposium on Foundations of Computer Science*, pages 2–13, 1992.

[Bab85]   L. Babai. Trading group theory for randomness. In *Proceedings of the 17th Annual ACM Symposium on Theory of Computing*, pages 421–429, 1985.

[BFL90]   L. Babai, L. Fortnow, and C. Lund. Nondeterministic exponential time has two–prover interactive protocols. In *Proceedings of the 31st IEEE Symposium on Foundations of Computer Science*, pages 16–25, 1990.

[BFLS91]   L. Babai, L. Fortnow, L. Levin, and M. Szegedy. Checking computations in polylogarithmic time. In *Proceedings of the 23rd Annual ACM Symposium on Theory of Computing*, pages 21–31, 1991.

[BH77]   L. Berman and J. Hartmanis. On isomorphism and density of NP and other complete sets. *SIAM Journal on Computing*, 6:305–322, 1977.

[BM88]   L. Babai and S. Moran. Arthur-Merlin games: a randomized proof system and a hierarchy of complexity classes. *Journal of Computer and System Sciences*, 34:254–276, 1988.

[BGKW88]   M. Ben-or, S. Goldwasser, J. Kilian, and A. Wigderson. Multi prover interactive proofs: How to remove intractability. In *Proceedings of the 20th Annual ACM Symposium on Theory of Computing*, pages 113–131, 1988.

[Coo71]   S. Cook. The complexity of theorem proving procedures. In *Proceedings of the 3rd Annual ACM Symposium on Theory of Computing*, pages 151–158, 1971.

[FGL$^+$91]   U. Feige, S. Goldwasser, L. Lovasz, S. Safra, and M. Szegedy. Approximating clique is alomst NP-complete. In *Proceedings of the 32nd IEEE Symposium on Foundations of Computer Science*, pages 2–12, 1991.

[FRS88]   L. Fortnow, J. Rompel, and M. Sipser. On the power of multi-prover interactive protocols. In *Proceedings of the 3rd Structure in Complexity Theory Conference*, pages 156–161, 1988.

[FS88]   L. Fortnow and M. Sipser. Are there interactive protocols for co-NP languages? *Information Processing Letters*, 28(5):249–251, 1988.

[Gö31]   K. Gödel. Über formal unentscheidbare Sätze der Principiua mathematica und verwandter Systeme. *Monatshefte für Mathematik und Physik*, 38:173–198, 1931.

[GJ79]   M. Garey and D. Johnson. *Computers and Intractability:A guide to the theory of NP-Completeness*. Freeman, 1979.

[GMR89]   S. Goldwasser, S. Micali, and C. Rackoff. The Knowledge complexity of interactive proof systems. *SIAM Journal on Computing*, 18:186–208, 1989.

[GMW86]   O. Goldreich, S. Micali, and A. Wigderson. Proofs that yield nothing but their validity and a methodology of cryptographic protocol design. In *Proceedings of the 27th IEEE Symposium on Foundations of Computer Science*, pages 174–187, 1986.

[Har89]   J. Hartmanis. Gödel, von Neumann and the P=?NP problem. *Bulletin of the EATCS*, 38:101–107, June 1989.

[HCRR90a]   J. Hartmanis, R. Chang, D. Ranjan, and P. Rohatgi. On IP=PSPACE and theorems with narrow proofs. *Bulletin of the EATCS*, 41:166–174, June 1990.

[HCRR90b]   J. Hartmanis, R. Chang, D. Ranjan, and P. Rohatgi. Structural Complexity Theory: Recent Surprises. In *Proceedings of SWAT 90*, pages 1–12. Lecture Notes in Computer Science #447, 1990.

[HLS65]   J. Hartmanis, P. Lewis, and R. Stearns. Hierarchies of memory limited computations. In *Proceedings of 6th IEEE Symposium on Switching Circuit Theory and Logical Design*, pages 179–190, 1965.

[HS65]   J. Hartmanis and R. Stearns. On the computational complexity of algorithms. *Trans. AMS*, 117:285–306, 1965.

[Imm89]   N. Immerman. Descriptive and computational complexity. In J. Hartmanis, editor, *Proceedings of Symposia in Applied Mathematics*, pages 75–91. AMS, 1989.

[Kad88]   J. Kadin. The polynomial time hierarchy collapses if the Boolean hierarchy collapses. *SIAM Journal on Computing*, 17(6):1263–1282, 1988.

[Kar72]   R. Karp. Reducibility among combinatorial problems. In R. Miller and J. Thatcher, editors, *Complexity of Computer Computations*, pages 85–103. Plenum Press, 1972.

[Lev73]   L. Levin. Universal'nyie perebornyie zadachi(universal search problems). *Problemy Peredachi Informatsii*, 9(3), 1973.

[LFKN90]   C. Lund, L. Fortnow, H. Karloff, and N. Nisan. Algebraic methods for interactive proof systems. In *Proceedings of the 31st IEEE Symposium on Foundations of Computer Science*, pages 2–10, 1990.

[LY93]   C. Lund and M. Yannakakis. On the hardness of approximating minimization problems. In *Proceedings of the 25th Annual ACM Symposium on Theory of Computing*, pages 286–293, 1993.

[Mah82]   S. Mahaney. Sparse complete sets for NP: Solution of a conjecture of Berman and Hartmanis. *Journal of Computer and System Sciences*, 25(2):130–143, 1982.

[OW90]   M. Ogiwara and O. Watanabe. On poynomial time bounded truth–table reducibility of NP to sparse sets. In *Proceedings of the 22nd Annual ACM Symposium on Theory of Computing*, pages 457–467, 1990.

[Roh92]   P. Rohatgi. Saving queries with randomness. In *Proceedings of the 7th Structure in Complexity Theory Conference*, pages 71–83, 1992.

[Roh94]   P. Rohatgi. *On Properties of Random Reductions*. PhD thesis, Cornell University, 1994. Available as Computer Science Department technical report TR 93–1386.

[Sha90]   A. Shamir. IP = PSPACE. In *Proceedings of the 31st IEEE Symposium on Foundations of Computer Science*, pages 11–15, 1990.

Advances in Graph Drawing[*]

Ashim Garg and Roberto Tamassia

Department of Computer Science
Brown University
Providence, RI 02912-1910
{ag,rt}@cs.brown.edu

Abstract. Graph drawing addresses the problem of constructing geometric representations of abstract graphs and networks. It is an emerging area of research that combines flavors of topological graph theory and computational geometry. The automatic generation of drawings of graphs has important applications in key computer technologies such as software engineering, database design, visual interfaces, and computer-aided-design. This paper surveys recent results of the authors on graph drawing and overviews various research trends in the area.

1 Introduction

Graph drawing addresses the problem of constructing geometric representations of abstract graphs and networks. It is an emerging area of research that combines flavors of topological graph theory and computational geometry. The automatic generation of drawings of graphs has important applications in key computer technologies such as software engineering, database design, visual interfaces, and computer-aided-design.

Research on graph drawing has been especially active in the last decade. The latest version of the annotated bibliography on graph drawing algorithms, by G. Di Battista, P. Eades, R. Tamassia and I.G. Tollis [18] lists more than 300 papers, and the *Graph Drawing '93* workshop [17] gathered 80 scientists from 19 countries.

In this paper we survey recent results of the authors on graph drawing and overview various research trends in the area. We do not aim at a comprehensive coverage of recent literature on the subject (see [18] for an up-to-date bibliography). Rather, we focus on specific problems related to the authors' own research, which are of theoretical and practical interest, and on which substantial progress has been recently made.

The rest of this paper is organized as follows. Basic definitions are given in Section 2. Section 3 discusses trade-offs between various properties of drawings. Section 4 deals with upward planarity testing. Techniques for dynamic graph drawing are presented in Section 5. Finally, work in progress on a new visual approach to graph drawing is discussed in Section 6.

---

[*] Research supported in part by the National Science Foundation under grant CCR-9007851, by the U.S. Army Research Office under grant DAAL03-91-G-0035, and by the Office of Naval Research and the Advanced Research Projects Agency under contract N00014-91-J-4052, ARPA order 8225.

2 Definitions

In this section we review basic graph drawing terminology. Various graphic standards have been proposed for the representation of graphs in the plane. Usually, vertices are represented by points, and each edge (u, v) is represented by a simple open Jordan curve joining the points associated with the vertices u and v. A drawing such that each edge is represented by a polygonal chain is a *polyline* drawing. There are two common special cases of this standard. A *straight-line* drawing maps each edge into a straight-line segment. An *orthogonal* drawing maps each edge into a chain of horizontal and vertical segments. A polyline drawing is a *grid* drawing if the vertices and the bends of the edges have integer coordinates. The *angular resolution* of a straight-line drawing is the smallest angle formed by two edges incident on the same vertex. This definition can be extended to polyline drawings by considering also the angles formed by the bends.

A drawing is *planar* if no two edges intersect. A drawing of a digraph is *upward* if every edge is monotonically nondecreasing in the y-direction. A digraph is *upward planar* if it admits a planar upward drawing. Planarity and acyclicity are necessary but not sufficient conditions for upward planarity.

A graph drawing algorithm reads as input a combinatorial description of a graph G, and produces as output a drawing of G according to a given graphic standard.

3 Trade-Offs in Graph Drawing

Many graph drawing problems can be formalized as multi-objective optimization problems. E.g., construct a drawing with minimum area and minimum number of crossings. Hence, trade-offs are inherent in graph drawing. Several interesting trade-offs have been recently discovered relating important properties of drawings, such as area, angular resolution, number of crossings, number of bends, and upwardness (for digraphs).

3.1 Upward Planar Drawings of Digraphs

Di Battista, Tamassia, and Tollis [23] study the problem of constructing upward planar polyline drawings with the the objective of achieving small area and small number of bends. Their results, summarized below, show that these two properties cannot be simultaneously attained, and that the area can become exponential if a straight-line drawing (i.e., a drawing the minimum number of bends) is required:

- There exists a family G_n of planar digraphs such that G_n has $2n + 2$ vertices, and a planar straight-line upward drawing of G_n has area $\Omega(2^n)$.
- Let G be an upward planar digraph with n vertices. A planar polyline upward grid drawing of G with $O(n^2)$ area, at most one bend per edge, and at most $2n - 5$ bends in total can be constructed in $O(n)$ time.

Note that if we relax the upwardness requirement, then area $O(n^2)$ is asymptotically optimal in the worst case, and an $O(n^2)$-area planar straight-line drawing can be constructed in $O(n)$ time [8, 14, 15, 48, 33].

3.2 Upward Planar Drawings of Trees

Garg, Goodrich, and Tamassia [27] study the aforementioned problem of constructing upward planar polyline drawings for the class of rooted trees (where the edges are directed towards the root). Again, they aim at achieving small area and small number of bends. Their results, summarized below, show that the area of the drawing depends on the restrictions on the number and type of bends, and on the possibility of changing the left-to-right order of the children.

- A bounded-degree rooted tree with n vertices admits a planar polyline upward grid drawing with $O(n)$ area, $O(n^\alpha)$ width, and $O(n^{1-\alpha})$ height (for any pre-specified constant α such that $0 < \alpha < 1$) that can be constructed in $O(n)$ time. Also, the above area bound is asymptotically optimal in the worst case.
- A binary tree with n vertices admits a planar orthogonal upward grid drawing with area $O(n \log \log n)$ that can be constructed in $O(n)$ time. Also, the above area bound is asymptotically optimal in the worst case.
- An ordered bounded-degree rooted tree with n vertices admits an $O(n \log n)$-area planar polyline upward grid drawing that preserves the left-to-right ordering of the children of each vertex. The drawing can be constructed in $O(n)$ time. Also, the above area bound is asymptotically optimal in the worst case.

Previously, Crescenzi, Di Battista, and Piperno [10] considered *strictly upward* grid drawings, where the vertices have integer coordinates, and the parent of a vertex has y-coordinate strictly greater than the ones of its children. They show that a rooted tree with n vertices, admits a strictly upward planar straight-line grid drawing with $O(n \log n)$ area, $O(n)$ width, and $O(\log n)$ height that can be constructed in $O(n)$ time. Also, they show that a complete binary tree or a Fibonacci tree with n vertices admits a planar straight-line strictly upward grid drawing with $O(n)$ area, and $O(\sqrt{n})$ width and height, that can be constructed in $O(n)$ time. The above area bounds are asymptotically optimal in the worst case.

Other interesting trade-offs arise if we relax the upwardness requirement. As shown in [39, 53], a binary tree with n vertices admits an $O(n)$-area planar orthogonal grid drawing. However, Brent and Kung [7] show that if the leaves of a complete binary tree with n vertices are constrained to be on the convex hull of the drawing, then the drawing needs $\Omega(n \log n)$ area.

3.3 Angular Resolution

Formann, Hagerup, Haralambides, Kaufmann, Leighton, Simvonis, Welzl, and Woeginger [26] study the angular resolution of (generally nonplanar) straight-line drawings of various classes of graphs. In particular, they show that a degree-d graph with n vertices has a straight-line drawing with angular resolution $\Omega(1/d^2)$ and area $O(d^6 n)$, and every degree-d planar graph with n vertices has a straight-line grid drawing with angular resolution $\Omega(1/d)$ and area $O(d^3 n)$.

Further work has been recently done on the angular resolution of planar drawings. Malitz and Papakostas [41] show that a degree-d planar graph admits a planar straight line drawing with angular resolution $\Omega(1/7^d)$. Kant [34] shows that a degree-d triconnected planar graph with n vertices admits a planar polyline grid

drawing with angular resolution $\Omega(1/d)$, $O(n^2)$ area and $O(n)$ bends, and that testing whether a biconnected planar graph admits a planar straight-line drawing with angular resolution greater than or equal to a given constant is NP-hard. Di Battista and Vismara [24] give an $O(n)$-time algorithm for testing whether a given assignment of angles to a maximal planar graph can be achieved in a planar straight-line drawing.

Garg and Tamassia [28] show a continuous trade-off between the area and the angular resolution of planar drawings:

- There exists a constant $c > 1$ and a family G_n^d of planar graphs such that G_n^d has n vertices and degree d, and any planar straight-line drawing of G_n^d with angular resolution ρ has area $\Omega(c^{\rho n})$.

In particular, the above result implies that there exist bounded-degree planar graphs that require exponential area in any planar straight-line drawing with bounded resolution.

3.4 Visibility Representations of Trees

The concept of *visibility* (see, e.g., [45, 43, 49]) plays an important role in computational geometry, and arises in art gallery problems, motion planning, and graphics. Given a set of disjoint objects in the plane (e.g., points, lines, rectangles), two objects of are said to be *1-visible* if they can be joined by a vertical segment that does not intersect any other object. A *1-visibility representation* of a graph maps the vertices to disjoint horizontal segments such that any two adjacent vertices are 1-visible. It is easy to show that a graph admits a 1-visibility representation only if it is planar. A 1-visibility representation of a digraph is said to be upward if for every edge (u, v), the vertex-segment of u is placed below the vertex-segment of v.

We distinguish three types of visibility representations. In a *weak* 1-visibility representation, adjacent vertices are associated with visible vertex-segments. However, visible vertex-segments are not necessarily associated with adjacent vertices. Tamassia and Tollis [51] and Rosenstiehl and Tarjan [47] show that every planar graph admits a weak 1-visibility representation. In *strong* and *ϵ-1-visibility* representations, two vertex-segments are 1-visible *if and only if* their associated vertices are adjacent. The two representation differ in that an ϵ-1-visibility representation allows open or semi-open vertex-segments (i.e., which may or may not contain one or both endpoints), while a strong 1-visibility representation allows only closed vertex-segments (i.e., with both endpoints). Tamassia and Tollis [51] and Wismath [54] characterize the class of graphs that admit an ϵ-1-visibility representation, and provide a linear-time algorithm for testing membership in this class. Tamassia and Tollis [51] also show that every 4-connected planar graph has a strong 1-visibility representation. However, Andreae [1] proves that deciding whether a graph admits a 1-strong visibility representation is NP-complete.

Algorithms for constructing in $O(n)$ time an $O(n^2)$-area weak 1-visibility representation of an n-vertex planar graph are given by Tamassia and Tollis [51] and by Rosenstiehl and Tarjan [47]. Kant [34] shows that a weak 1-visibility representation of an n-vertex planar graph with area at most $(\lfloor \frac{3}{2} n \rfloor - 3) \times (n - 1)$ can be constructed in $O(n)$ time.

In *2-visibility representations* of graphs the vertices are represented by disjoint isothetic rectangles in the plane, and adjacent vertices are associated with rectangles that are visible in the vertical or horizontal direction. As for the 1-visibility case, we distinguish *weak*, *strong*, and ϵ-2-visibility representations. Kirkpatrick and Wismath [38] prove that every planar graph admits an ϵ-2-visibility representation.

Kant, Liotta, Tamassia, and Tollis [35] study the area requirement of 1- and 2-visibility representations of trees. Their results show how the area requirement varies with the specific type of visibility representation.

The following definitions are needed: Let T be a free tree, and let v be a vertex of T. A *subtree* of v in T is a subtree of the rooted tree obtained by rooting T at v. Let $T_1, T_2, \cdots, T_d$ be the subtrees of v sorted by nonincreasing height (i.e., $h(T_1) \geq h(T_2) \geq \cdots \geq h(T_d)$). We call $h(T_k)$ the *k-th height* of v. A subtree of v with height equal to the k-th height of v is called a *k-tallest subtree* of v. If T has degree at least 3, we call *critical height* of T the maximum third-height of any vertex of T.

- Let T be a rooted tree with n vertices, ℓ leaves and height h. The area required by an upward 1-visibility representation of T is $\Omega(\ell \cdot h)$. Also, an upward 1-visibility representation of T with $O(\ell \cdot h)$ area can be computed in $O(n)$ time.
- Let T be a free tree with n vertices, ℓ leaves, and critical height h^*. The area required by a strong or ϵ-1-visibility representation of T is $\Omega(n + \ell \cdot h^*)$. Also, a strong or ϵ-1-visibility representation of T with $O(n+\ell\cdot h^*)$ area can be computed in $O(n)$ time.
- Let T be a free tree with n vertices and ℓ leaves. The area required by a strong or ϵ-2-visibility representation of T is $\Omega(n \cdot \ell)$. Also, a strong or ϵ-2-visibility representation of T with $O(n \cdot \ell)$ area can be computed in $O(n)$ time.

4 Upward Planarity Testing

Testing upward planarity and constructing upward planar drawings of digraphs is important for displaying hierarchical structures, such order diagrams, subroutine-call graphs, is-a hierarchies, and organization charts. While planarity testing can be done in linear time [30, 40, 6, 16], the complexity of upward planarity testing has been until very recently an open problem.

Combinatorial results on upward planarity of covering digraphs of lattices were first given in [37, 44]. The interplay between upward planarity and ordered sets is surveyed in [46]. Lempel, Even, and Cederbaum [40] relate the planarity of biconnected undirected graphs to the upward planarity of *st*-digraphs.

A combinatorial characterization of upward planar digraphs is provided in [36, 22]: namely, a digraph is upward planar if and only if it is a subgraph of a planar *st-digraph*. Di Battista, Liu, and Rival [21] shown that every planar bipartite digraph is upward planar. Bertolazzi, Di Battista, Liotta, and Mannino [3, 4] give a polynomial-time algorithm for testing upward planarity of triconnected digraphs and digraphs with a fixed embedding. Concerning single-source digraphs, Thomassen [52] characterizes upward planarity in terms of forbidden circuits. Hutton and Lubiw [31] combine Thomassen's characterization with a decomposition scheme to test upward planarity of an n-vertex single-source digraph in $O(n^2)$ time.

Bertolazzi, Di Battista, Mannino, and Tamassia [5] give optimal algorithms for upward planarity testing of single-source digraphs:

- Upward planarity testing of a single-source digraph with n vertices can be done in $O(n)$ time, and in $O(\log n)$ time on a CRCW PRAM with $n \log \log n / \log n$ processors.

Garg and Tamassia [29] have recently settled the open question of the complexity of upward planarity testing:

- Upward planarity testing is NP-complete.

5 Dynamic Graph Drawing

The motivation for investigating dynamic graph drawing algorithms arises when very large graphs need to be visualized in a dynamic environment, where vertices and edges are inserted and deleted, and subgraphs are displayed. Several graph manipulation systems allow the user to interactively modify a graph; hence, techniques that support fast restructuring of the drawing are very useful. Also, it is important that the dynamic drawing algorithm does not alter drastically the structure of the drawing after a local modification of the graph. In fact, human interaction requires a "smooth" evolution of the drawing.

Cohen, Di Battista, Tamassia, Tollis, and Bertolazzi [9] present dynamic algorithms for drawing planar graphs under a variety of drawing standards. They consider straight-line, polyline, grid, upward, and visibility drawings together with aesthetic criteria that are important for readability, such as the display of planarity, symmetry, and reachability. Also, they provide techniques that are especially tailored for important subclasses of planar graphs such as trees and series-parallel digraphs. Their dynamic drawing algorithms have the important property of performing "smooth updates" of the drawing. Their contributions are summarized as follows:

- A model for dynamic graph drawing is presented that uses a "query/update" setting, where an *implicit* representation of the drawing of a graph is maintained such that the following operations can be efficiently performed: *Drawing queries* that return the drawing of a subgraph consistent with the overall drawing of the graph. *Window queries* that return the portion of the drawing inside a query rectangle. *Point-location queries* in the subdivision of the plane induced by the drawing. *Update operations*, e.g., insertion and deletion of vertices and edges or replacement of an edge by a graph, which modify the implicit representation of the drawing accordingly.
- There exists a dynamic algorithm for maintaining a planar upward straight-line $O(n^2)$-area grid drawing of a rooted tree with n vertices. The memory space used is $O(n)$. Updates and point-location queries take $O(\log n)$ time. Drawing queries take time $O(k + \log n)$ for a subtree and $O(k \log n)$ for an arbitrary subgraph, and window queries take time $O(k \log n)$, where k is the output size.

- There exists a dynamic algorithm for maintaining a planar upward straight-line $O(n^2)$-area grid drawing a of an n-vertex series-parallel digraph. The memory space used is $O(n)$, updates take $O(\log n)$ time, drawing queries take time $O(k + \log n)$ for a series-parallel subgraph and $O(k \log n)$ for an arbitrary subgraph, point location queries take $O(\log n)$ time, and window queries take $O(k \log^2 n)$ time, where k is the output size.
- There exist dynamic algorithms for maintaining various types of planar drawings of a planar st-digraph with n vertices, including polyline upward drawings, and visibility representations. The drawings occupy $O(n^2)$ area. The algorithms use $O(n)$ memory space and support updates in $O(\log n)$ time.
- There exist semi-dynamic algorithms for maintaining various types of planar drawings of a biconnected planar graph with n vertices, including polyline drawings and visibility representations. The algorithms uses $O(n)$ memory space and support insertions in $O(\log n)$ amortized time (worst-case for insertions that preserve the embedding). Drawing queries take $O(k \log n)$ time, where k is the output size.

6   A Visual Approach to Graph Drawing

Work in graph drawing has traditionally focused on *algorithmic* approaches, where the drawing of the graph is generated according to a prespecified set of general rules or aesthetic criteria (such as planarity or area minimization) that are embodied in an algorithm. Perhaps the most sophisticated graph drawing system based on the algorithmic approach is *Diagram Server* [2, 20, 19], which maintains a large database of graph drawing algorithms and is able to select the one best suited to the needs of the user.

The algorithmic approach is computationally efficient, however, it does not naturally support *constraints*, i.e., requirements that the user may want to impose on the drawing of a *specific* graph (e.g., clustering or aligning a given set of vertices). Previous work by Tamassia *et al.* [50] has shown the importance of satisfying constraints in graph drawing systems, and has demonstrated that a limited constraint satisfaction capability can be added to an existing drawing algorithm.

Recently, several attempts have been made at developing languages for the specification of constraints and at devising techniques for graph drawing based on the resolution of systems of constraints (see, e.g., [32, 42]). Current constraint-based systems have three major drawbacks: (*i*) The specification of constraints is made through a detailed enumeration of facts from a fixed set of predicates, expressed in Prolog or with a set-theoretic notation. (*ii*) Natural requirements, such as planarity, need complicated constraints to be expressed. (*iii*) General constraint-solving systems are computationally inefficient. The above discussion indicates the need for a language to specify constraints that reconciles expressiveness with efficiency.

Cruz, Tamassia and Van Hentenryck [13] are developing a new technique for the visual specification of constraints in graph drawing systems. Visual languages can provide a natural and user-friendly way to express the layout of a graph. For this purpose, they plan to design a variation of DOODLE, a visual language for the specification of the display of facts in an object-oriented database [11, 12].

Recent work by Eades and Lin [25] has similar objectives, but is not based on a visual specification.

References

1. T. Andreae. Some results on visibility graphs. *Discrete Applied Mathematics*, 40:5–17, 1992.

2. M. Beccaria, P. Bertolazzi, G. Di Battista, and G. Liotta. A tailorable and extensible automatic layout facility. In *Proc. IEEE Workshop on Visual Languages (VL'91)*, pages 68–73, 1991.

3. P. Bertolazzi and G. Di Battista. On upward drawing testing of triconnected digraphs. In *Proc. 7th Annu. ACM Sympos. Comput. Geom.*, pages 272–280, 1991.

4. P. Bertolazzi, G. Di Battista, G. Liotta, and C. Mannino. Upward drawings of triconnected digraphs. *Algorithmica*, to appear.

5. P. Bertolazzi, G. Di Battista, C. Mannino, and R. Tamassia. Optimal upward planarity testing of single-source digraphs. In *1st Annual European Symposium on Algorithms (ESA '93)*, Lecture Notes in Computer Science. Springer-Verlag, 1993.

6. K. Booth and G. Lueker. Testing for the consecutive ones property interval graphs and graph planarity using PQ-tree algorithms. *J. Comput. Syst. Sci.*, 13:335–379, 1976.

7. R.P. Brent and H.T. Kung. On the area of binary tree layouts. *Information Processing Letters*, 11:521–534, 1980.

8. M. Chrobak and T. H. Payne. A linear time algorithm for drawing a planar graph on a grid. Technical Report UCR-CS-90-2, Dept. of Math. and Comput. Sci., Univ. California Riverside, 1990.

9. R. F. Cohen, G. Di Battista, R. Tamassia, I. G. Tollis, and P. Bertolazzi. A framework for dynamic graph drawing. In *Proc. 8th Annu. ACM Sympos. Comput. Geom.*, pages 261–270, 1992.

10. P. Crescenzi, G. Di Battista, and A. Piperno. A note on optimal area algorithms for upward drawings of binary trees. *Computational Geometry: Theory and Applications*, 2:187–200, 1992.

11. I. F. Cruz. DOODLE: A visual language for object-oriented databases. In *Proc. ACM SIGMOD*, pages 71–80, 1992.

12. I. F. Cruz. Using a visual constraint language for data display specification. In P. C. Kanellakis, J.-L. Lassez, and V. Saraswat, editors, *First Workshop on Principles and Practice of Constraint Programming*, Newport, RI, April 1993.

13. I. F. Cruz, R. Tamassia, and P. Van Hentenryk. A visual approach to graph drawing. In *Graph Drawing '93 (Proc. ALCOM Workshop on Graph Drawing)*, Paris, France, September 1993.

14. H. de Fraysseix, J. Pach, and R. Pollack. Small sets supporting Fary embeddings of planar graphs. In *Proc. 20th Annu. ACM Sympos. Theory Comput.*, pages 426–433, 1988.

15. H. de Fraysseix, J. Pach, and R. Pollack. How to draw a planar graph on a grid. *Combinatorica*, 10:41–51, 1990.

16. H. de Fraysseix and P. Rosenstiehl. A depth-first-search characterization of planarity. *Annals of Discrete Mathematics*, 13:75–80, 1982.

17. G. Di Battista, P. Eades, H. de Fraysseix, P. Rosenstiehl, and R. Tamassia. *Graph Drawing '93 (Proc. ALCOM Int. Workshop on Graph Drawing)*. 1993. Available via anonymous ftp from wilma.cs.brown.edu, /pub/papers/compgeo/gd93-v2.tex.Z.

18. G. Di Battista, P. Eades, R. Tamassia, and I. G. Tollis. Algorithms for drawing graphs: an annotated bibliography. Preprint, Dept. Comput. Sci., Brown Univ., Providence,

RI, November 1993. To appear in *Comput. Geom. Theory Appl.* Preliminary version available via anonymous ftp from vilma.cs.brown.edu, gdbiblio.tex.Z and gdbiblio.ps.Z in /pub/papers/compgeo.

19. G. Di Battista, A. Giammarco, G. Santucci, and R. Tamassia. The architecture of diagram server. In *Proc. IEEE Workshop on Visual Languages (VL'90)*, pages 60–65, 1990.

20. G. Di Battista, G. Liotta, M. Strani, and F. Vargiu. Diagram server. In *Advanced Visual Interfaces (Proceedings of AVI '92)*, volume 36 of *World Scientific Series in Computer Science*, pages 415–417, 1992.

21. G. Di Battista, W. P. Liu, and I. Rival. Bipartite graphs upward drawings and planarity. *Inform. Process. Lett.*, 36:317–322, 1990.

22. G. Di Battista and R. Tamassia. Algorithms for plane representations of acyclic digraphs. *Theoret. Comput. Sci.*, 61:175–198, 1988.

23. G. Di Battista, R. Tamassia, and I. G. Tollis. Area requirement and symmetry display of planar upward drawings. *Discrete Comput. Geom.*, 7:381–401, 1992.

24. G. Di Battista and L. Vismara. Angles of planar triangular graphs. In *Proc. 25th Annu. ACM Sympos. Theory Comput. (STOC 93)*, pages 431–437, 1993.

25. P. Eades and T. Lin. Algorithmic and declarative approaches to aesthetic layout. In *Graph Drawing '93 (Proc. ALCOM Workshop on Graph Drawing)*, Paris, France, September 1993.

26. M. Formann, T. Hagerup, J. Haralambides, M. Kaufmann, F. T. Leighton, A. Simvonis, E. Welzl, and G. Woeginger. Drawing graphs in the plane with high resolution. In *Proc. 91th Annu. IEEE Sympos. Found. Comput. Sci.*, pages 86–95, 1990.

27. A. Garg, M. T. Goodrich, and R. Tamassia. Area-efficient upward tree drawings. In *Proc. 9th Annu. ACM Sympos. Comput. Geom.*, pages 359–368, 1993.

28. A. Garg and R. Tamassia. Angular resolution of planar drawings. Technical report, Brown Univ., Dept. of Computer Science, 1993.

29. A. Garg and R. Tamassia. On the complexity of upward planarity testing. Technical report, Brown Univ., Dept. of Computer Science, 1993.

30. J. Hopcroft and R. E. Tarjan. Efficient planarity testing. *J. ACM*, 21(4):549–568, 1974.

31. M. D. Hutton and A. Lubiw. Upward planar drawing of single source acyclic digraphs. In *Proc. 2nd ACM-SIAM Sympos. Discrete Algorithms*, pages 203–211, 1991.

32. T. Kamada. *Visualizing Abstract Objects and Relations*. World Scientific Series in Computer Science, 1989.

33. G. Kant. Drawing planar graphs using the lmc-ordering. In *Proc. 33th Annu. IEEE Sympos. Found. Comput. Sci.*, pages 101–110, 1992.

34. G. Kant. A more compact visibility representation. In *Proc. 19th Internat. Workshop Graph-Theoret. Concepts Comput. Sci. (WG'93)*, 1993.

35. G. Kant, G. Liotta, R. Tamassia, and I. Tollis. Area requirements of visibility representations of trees. In *Proc. 5th Canad. Conf. Comput. Geom.*, pages 192–197, Waterloo, Canada, 1993.

36. D. Kelly. Fundamentals of planar ordered sets. *Discrete Math.*, 63:197–216, 1987.

37. D. Kelly and I. Rival. Planar lattices. *Canad. J. Math.*, 27(3):636–665, 1975.

38. D. G. Kirkpatrick and S. K. Wismath. Weighted visibility graphs of bars and related flow problems. In *Proc. 1st Workshop Algorithms Data Struct.*, volume 382 of *Lecture Notes in Computer Science*, pages 325–334. Springer-Verlag, 1989.

39. C. E. Leiserson. Area-efficient graph layouts (for VLSI). In *Proc. 21st Annu. IEEE Sympos. Found. Comput. Sci.*, pages 270–281, 1980.

40. A. Lempel, S. Even, and I. Cederbaum. An algorithm for planarity testing of graphs. In *Theory of Graphs: Internat. Symposium (Rome 1966)*, pages 215–232, New York, 1967. Gordon and Breach.

41. S. Malitz and A. Papakostas. On the angular resolution of planar graphs. In *Proc. 24th Annu. ACM Sympos. Theory Comput.*, pages 527–538, 1992.

42. J. Marks. A formal specification for network diagrams that facilitates automated design. *Journal of Visual Languages and Computing*, 2:395–414, 1991.

43. J. O'Rourke. *Art Gallery Theorems and Algorithms*. Oxford University Press, New York, NY, 1987.

44. C. Platt. Planar lattices and planar graphs. *J. Combin. Theory Ser. B*, 21:30–39, 1976.

45. F. P. Preparata and M. I. Shamos. *Computational Geometry: an Introduction*. Springer-Verlag, New York, NY, 1985.

46. I. Rival. Graphical data structures for ordered sets. In I. Rival, editor, *Algorithms and Order*, pages 3–31. Kluwer Academic Publishers, 1989.

47. P. Rosenstiehl and R. E. Tarjan. Rectilinear planar layouts and bipolar orientations of planar graphs. *Discrete Comput. Geom.*, 1(4):343–353, 1986.

48. W. Schnyder. Embedding planar graphs on the grid. In *Proc. 1st ACM-SIAM Sympos. Discrete Algorithms*, pages 138–148, 1990.

49. T. C. Shermer. Recent results in art galleries. *Proc. IEEE*, 80(9):1384–1399, September 1992.

50. R. Tamassia, G. Di Battista, and C. Batini. Automatic graph drawing and readability of diagrams. *IEEE Trans. Syst. Man Cybern.*, SMC-18(1):61–79, 1988.

51. R. Tamassia and I. G. Tollis. A unified approach to visibility representations of planar graphs. *Discrete Comput. Geom.*, 1(4):321–341, 1986.

52. C. Thomassen. Planar acyclic oriented graphs. *Order*, 5(4):349–361, 1989.

53. L. Valiant. Universality considerations in VLSI circuits. *IEEE Trans. Comput.*, C-30(2):135–140, 1981.

54. S. K. Wismath. Characterizing bar line-of-sight graphs. In *Proc. 1st Annu. ACM Sympos. Comput. Geom.*, pages 147–152, 1985.

On a parallel-algorithms method for string matching problems (overview)

Suleyman Cenk Sahinalp *   *Uzi Vishkin* [†]

Abstract

Suffix trees are the main data-structure in string matching algorithmics. There are several serial algorithms for suffix tree construction which run in linear time, but the number of operations in the only parallel algorithm available, due to Apostolico, Iliopoulos, Landau, Schieber and Vishkin, is proportional to $n \log n$. The algorithm is based on labeling substrings, similar to a classical serial algorithm, with the same operations bound, by Karp, Miller and Rosenberg. We show how to break symmetries that occur in the process of assigning labels using the Deterministic Coin Tossing (DCT) technique, and thereby reduce the number of labeled substrings to linear.

---

[*]Department of Computer Science, University of Maryland at College Park

[†]University of Maryland Institute for Advanced Computer Studies, College Park, MD 20742; and Dept. of Computer Science, Tel Aviv University, Tel Aviv, Israel; Partially supported by NSF grants CCR-8906949 and CCR-9111348.

1 Introduction

Suffix trees are apparently the single most important data-structure in the area of string matching.

We present a parallel method for constructing the suffix tree T of a string $S = s_1 \ldots s_n$ of n symbols, with s_n being a special symbol \$ that appears nowhere else in S. We use A to denote the *alphabet* of S. The suffix tree T associated with S is a rooted tree with n leaves such that:

(1) Each path from the root to a leaf of T represents a different suffix of S.

(2) Each edge of T represents a nonempty substring of S.

(3) Each nonleaf node of T, except the root, must have at least two children.

(4) The substrings represented by two sibling edges must begin with different characters.

An example of a suffix tree is given in Figure 1.

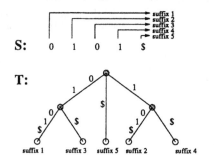

Figure 1: Suffix tree T of string $S = 0\ 1\ 0\ 1\ \$$

Serial algorithms for suffix tree construction were given in [KMR72], [We73], and [Mc76]. The two latter algorithms achieve a linear running time for an alphabet whose size is constant.

A parallel algorithm was given in [AILSV88].

A Symmetry Breaking Challenge: As in the algorithm of [KMR72] work complexity of the above mentioned parallel algorithm is $O(n \log n)$. The approach of [KMR72] and [AILSV88] does not lend itself to linear work for the following reason: As these algorithms progress, they label all $n - 1$ substrings of size 2, then all $n - 3$ substrings of size 4, and in general all $n - 2^i + 1$ substrings of size 2^i ($1 \leq i \leq \log n$). This results in a number of labels which is proportional to $n \log n$ and this dictates the work complexity. The extra logarithmic factor in the label-count is due to the increasing redundance among these substrings (because of the overlaps), as they become longer. The problem is that there has been no consistent way for selecting only one among a subset of overlapping substrings, since they all "look-alike". The main new idea of this paper is in introducing a solution to this *symmetry breaking* *problem.* Our most interesting concrete result is in being able to build a suffix tree using only a linear number of labels.

The general area of string matching has been enriched by parallel methods that enabled new serial algorithms as in this paper. Previous examples include [Ga85], [Vi85], and [Vi91]. The new method is also relevant for sequence analysis in compressed data, since it allows for consistent compression of data. This can be done in the context of parallel or serial algorithms. Applications of the new method for data compression will be discussed in the full version.

The method described in this paper leads to several incomparable complexity results as described in [SV93]. We quote here only one which can be derived with reasonable effort from our description. For an alphabet whose size is polynomial in n, the method gives an $O(n \log^* n)$ work algorithms and $O(n^\epsilon)$ time for any constant $0 < \epsilon \le 1$.

2 The Algorithm

2.1 High-level Description

The algorithm works in two stages. In the **first stage** we attach **labels** to various substrings of S, recognizing some identities. This is done in iterations. In iteration 1, S is partitioned into blocks of size 2 or 3 characters. Each block is labeled with a number between 1 and n, in a way which satisfies the following two **consistency properties:**

Partition-consistency (we state this property informally) Let X_i be a "long enough" substring of S (starting) at location i and let X_j be a substring at location j, which is equal to X_i; then, with the exception of some margins, X_i and X_j will be partitioned in the same way.

Label-consistency All blocks consisting of the same string of characters will get the same label.

An example of consistent partitioning and consistent labeling is given in Figure 2.

So, iteration 1 "shrinks" $S = S(0)$, into a new string $S(1)$, reducing its length by a factor of at least two. Subsequent iterations apply the same procedure. Iteration i, $i = 2, 3, \ldots$ shrinks string $S(i-1)$ into string $S(i)$ satisfying similar partition-consistency and label-consistency properties. The size of strings $S(i)$ will be at most $n/2^i$.

The **second stage** is devoted to constructing the suffix tree of S in iterations. The input for the last iteration is $T(1)$, which is the suffix tree of (some kind of) labels which are derived from $S(1)$. The last iteration constructs the suffix tree of $S(0) = S$. The i'th-prior-to-the-last iteration constructs the suffix tree $T(i)$ (the suffix tree of labels derived from $S(i)$), by using $T(i+1)$.

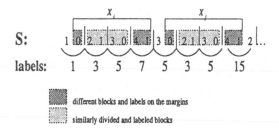

Figure 2: Consistent partitioning, margins, and consistent labeling

2.2 First Stage

Let R, a string of characters (of size m), be the input for an iteration of the **first stage**. The iteration partitions R into blocks of size 2 or 3 and labels each block. The problem is how to do it to satisfy partition-consistency (defined formally later) and label-consistency. We first describe the main steps of an iteration, and then give a detailed description.

The main steps of an iteration:

Each character r_i checks if it is in a substring of length 2, or more, of a single repeated character. If yes (i.e, $r_i = r_{i-1}$ or $r_i = r_{i+1}$), it uses procedure LENGTH-BASED partitioning (or, procedure LENGTH-BASED for short) for obtaining the block partitioning. If no, it uses procedure CONTENT-BASED partitioning (or, procedure CONTENT-BASED for short) for obtaining the block partitioning. (*Comment:* The case were some character r_i is not part of a substring of a single repeated character but its previous character r_{i-1} as well its next character r_{i+1} are both part of such substrings, need a separate treatment which is suppressed in this presentation.)

The reader is asked to be alert to the following: in order to better convey the main ideas of our algorithm, we **systematically suppressed, or deferred** dealing with the important case of substrings consisting of a single repeated character. Dealing with this case is relatively simpler, and it does not change the essentials of most of the new techniques presented in this paper.

We start with describing procedure CONTENT-BASED. Consider a substring $R_D = r_\alpha \ldots r_\beta$, where $r_i \neq r_{i+1}$, for $\alpha \leq i < \beta$. Namely, we do not allow a substring of the form aa. The main idea behind the *partitioning procedure* below is the use of the *deterministic coin tossing technique* of Cole and Vishkin [CV86a] for dividing R_D into blocks.

(*Comment.* Our use of the deterministic coin tossing technique is novel. We will use it for deriving "signatures" of strings, mapping similar substrings to the same signature. The only previous paper which made use of this technique for producing signatures is apparently by Mehlhorn, Sundar and Uhrig [MSU94]. They limited these signatures to compare full strings, and did not have to deal either with consistency relative to substrings, or to consider assigning signatures for strings of repeated characters (i.e. in the form aaa...).)

1. Put a divider to the left of r_α and to the right of r_β. Each instruction for putting a divider below should be augmented with the following *caveat*: we actually put the new divider only if it does not create a block consisting of a single character.

2. Put a divider to the right of $r_{\alpha+1}$. Put a divider to the left of $r_{\beta-1}$.

3. For each character r_i of R_D compute tag_i, the index of the least significant bit of r_i which is different than r_{i+1} in binary representation. If r_{i+1} does not exist (for now, this can happen only if $i = m$), set $tag_i := 0$.

4. For each character r_i of R_D, compare tag_i with tag_{i+1} and tag_{i-1}. If any of them does not exist (for now this can happen if $i = 1$, or $i = m$), take the non-existing value to be 0. For all "strict local maxima" (i.e., $tag_i > tag_{i+1}$ and $tag_i > tag_{i-1}$) put a block divider between r_i and r_{i+1}. For all "weakly local maxima" (i.e. $tag_i \geq tag_{i+1}$, and $tag_i \geq tag_{i-1}$), put a block divider between r_i and r_{i+1}, if bit $diff_i$ of r_i is 1.

5. For each substring of R_D, which lies between two dividers, do the following. If the substring has ≤ 3 elements those elements quit. Otherwise, for each character r_i, replace character r_i by tag_i, and recursively apply the CONTENT-BASED procedure separately to each substring (which lies between two dividers).

Example Let $R = \ldots 3\ 3\ 8\ 4\ 2\ 1\ 2\ 4\ 8\ 4\ 8\ 4\ 8\ 8 \ldots$. Typically, we will apply the CONTENT-BASED procedure to a longest substring which satisfies the input conditions which in this case will be:
$R_D = 8\ 4\ 2\ 1\ 2\ 4\ 8\ 4\ 8\ 4$. After applying steps 1 and 2 we have:
$|8\ 4|2\ 1\ 2\ 4\ 8\ 4|8\ 4|$. In binary representation R_D becomes:
1000 0100|0010 0001 0010 0100 1000 0100|1000 0100 , and the corresponding *tag* values are:
3 2 1 1 2 3 3 3 3 0 . Hence in the first round of CONTENT-BASED, we partition R_D as follows:
8 4|2 1 2 4|8 4|8 4 . Then we apply CONTENT-BASED procedure again to get:
8 4|2 1|2 4|8 4|8 4 , as the final partitioning of R_D.

Lemma 1.1 (the Block-partitioning Lemma): The above partitioning procedure divides R_D into blocks of size 2 or 3.

Proof: The Lemma follows from known facts about the *deterministic coin tossing technique* .

Lemma 1.2 (the Consistency Lemma): Let R_D be a substring of R in which no two successive characters are identical, and let R'_D be another substring of R which is identical with R_D. The CONTENT-BASED procedure partitions R_D and R'_D so that all but (at most) $\log^* n + 1$ blocks in the right margin and (at most) $\log^* n + 1$ blocks in the left margin are identical.

Proof: The lemma follows by a simple induction on the recursive call of CONTENT-BASED on R_D and R'_D. Designating a character as a local minima in call number p (which determines the location of a block divider) depends on at most $2p+3$ neighboring characters to its left and $2p+3$ neighboring characters to its right and we only have $\log^* n$ successive recursive calls for CONTENT-BASED.

Next, we proceed to consider a substring of a single repeated character $R_R = (r_{j+1} =$

$\ldots = r_{j+k})$ which cannot be further extended. That is, $r_j \neq r_{j+1} = r_{j+2} \ldots = r_{j+k} \neq r_{j+k+1}$.

We describe procedure LENGTH-BASED for obtaining a block-partitioning of R_R in Appendix 1. By applying these procedures one after another, we obtain a complete partitioning of R.

Now, we can attach labels to blocks preserving label-consistency, and finish the iteration.

Each block is labeled with a number between 1 and m to satisfy label-consistency. This can be done in a CRCW PRAM in $O(1)$ time with $O(m)$ work in two substeps. In the first substep we attach labels to all blocks of size 2, as well as to the first two characters of all blocks of size 3. This is done using an $m \times m$ array L. For each such two-character substring ij, where $1 \leq i, j \leq m$, the location of i is entered into entry $L(i, j)$. This is similar to [AILSV88]. In the second substep we attach labels to all block of size 3. For each three-character substring ijk, we obtain its label by applying the first substep to fk where f denotes the label of ij as computed in the first substep. The space complexity is $O(m^2)$. This can be reduced to $O(m^{1+\epsilon})$, for any fixed $\epsilon > 0$, as in [AILSV88]. Getting linear space using hashing (and thereby entering randomization), as suggested in [MV91], is also a possibility.

An example for block partitioning and labeling is given in Figure 3.

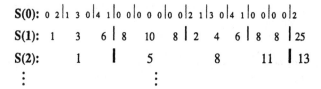

Figure 3: Block division and labeling for successive iterations

2.3   Second Stage

Before starting to describe the **Second Stage**, we mention the issues which will be suppressed in this presentation: (1) Data structures for representing the suffix tree and keeping the labels of substrings. (2) Processor allocation issues. (3) The case of substrings of a repeated character; so, for following our presentation the reader should assume that we never get two successive identical characters in any of the $S(k)$ sequences (for $k = 1, 2, \ldots,$).

Each iteration of the first stage provides a partition of the string into blocks and assigns labels to these blocks. the partitions in the first stage into k-substrings and the k-labels provided some limited similarities among substrings, as characterized by Lemma 1.2. For the construction of suffix trees later, it will be helpful to first develop another system of substrings of S (instead of the k-substrings) called **cores**, and label them with core **names**.

Any k-substring S_k induces another substring of the input which is called a k-**core**. Since

S_k is a substring of its k-core we say that S_k **spans** its k-core. An example of a core is given in Figure 4.

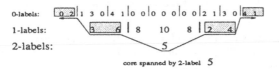

O-labels: | 0 2 | 1 3 0 | 4 1 0 0 0 0 0 0 0 0 2 1 3 0 | 4 1 |

1-labels: | 3 6 | 8 10 8 | 2 4 |

2-labels: 5

core spanned by 2-label 5

Figure 4: An example of a core for S in Fig.3; for illustration purposes, assume $2\log^* n + 3 = 2$ (which is not possible)

Given a substring S_k, we show how to **extend** it to the left and to the right to obtain a k-core. S_k is in the middle. To its left there will be $2\log^* n + 3$ strings which are $(k-1)$-substrings. To their left there will be will be again $2\log^* n + 3$ strings which are $(k-2)$-substrings, and so on till finally there will be $2\log^* n + 3$ which are (0)-substrings (they are, in fact, singletons). Finally, a k-core has also a symmetric "staircase" to the right of S_k. So, a k-core is a "double staircase" as illustrated in Figure 5.

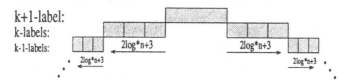

k+1-label:
k-labels:
k-1-labels:

$2\log^* n+3$ $2\log^* n+3$

$2\log^* n+3$ $2\log^* n+3$

Figure 5: A k-substring is extended by $2\log^* n + 3$, $k-1$-substrings followed by $2\log^* n + 3$, $k-2$-substrings... towards both left and right to obtain a k-core

The suffix (of the original pattern) which begins at the leftmost character of a k-core is called the **suffix of that core**, and is referred to as a k-**suffix**. (*Comment*: Each $(k+1)$ suffix is also a k-suffix. In other words, the suffix of every $(k+1)$-core is suffix of a k-core, since the leftmost character of the $(k+1)$-core is also the leftmost character of the k-core.

STEP 2.1 Corresponding to each iteration $k = 1, 2, \ldots$ of the **first stage**, do the following. For each k-substring, compute the k-core it spans and label each such core with a core name, which is called a k-name. Each k-core is actually a concatenation (with overlaps) of $4\log^* n + 7$, or $4\log^* n + 6$, $(k-1)$-cores; and the name of the k-core is derived from the names of these $(k-1)$-cores. A key point in deriving the k-names is that this is done **consistently**; that is, if the $k-1$-names form an identical sequence elsewhere then both k-cores will get the same k-name.

STEP 2.2 The suffix tree is constructed iteratively. In iteration k of the **Second Stage**, the suffix tree $T(k)$ of k-cores using k-names is built. This suffix tree has limited resolution, as some possible identical prefixes between two $(k+1)$-cores cannot be precisely expressed with $(k+1)$-names. **Iteration k builds $T(k)$ from $T(k+1)$ by improving the resolution with a more dense set of cores which are shorter.** Note that the iterations are numbered backwards, where the final iteration (which is iteration 1) gives the desired suffix tree.

The main ideas in the Second Stage We need to do several things in order to build $T(k)$ from $T(k+1)$.

1. **Get tree $T(k)_0$.** This is still the suffix tree of the $(k+1)$-suffixes (as $T(k+1)$), but it uses k-names. Procedure REFINE will derive tree $T(k)_0$ from tree $T(k+1)$.

Procedure REFINE will work as follows. Some $(k+1)$-cores represented in $T(k+1)$ are replaced with a concatenation (with overlaps) of the $4\log^* n + 9$, or $4\log^* n + 8$, k-cores that form them. In parallel for every edge of $T(k+1)$ advance step-by-step through its first $4\log^* n + 8$ (or $4\log^* n + 7$) k-cores, merging identical names into a single edge.

Observation: The common prefixes of k-cores between any two sibling edges in $T(k+1)$ can not be longer than $4\log^* n + 8$ (or $4\log^* n + 7$) as the definition of the $k+1$-core implies; hence the procedure REFINE will discover all similarities between $k+1$-cores in terms of k-cores in at most $4\log^* n + 8$ steps.

2. **Get tree $T(k)_1$.** For each $(k+1)$-core do the following. The $(k+1)$-core is spanned by some $(k+1)$-substring S_{k+1}. To the left of S_{k+1}, there will be $2\log^* n + 3$ strings which are k-substrings. Consider the next k-substring. An example is given in Figure 6. Pick the

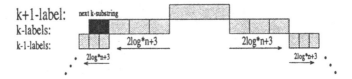

Figure 6: An illustration of the next k-substring of a $k+1$-core

k-core that it spans. Given all the k-cores that were picked, we build the suffix tree of their suffixes using k-names and denote it $T(k)_1$. This is done as follows. Divide the k-cores into equivalence classes, and let us focus on one of these equivalence classes, to be denoted Q. In $T(k)_1$, the suffixes of the cores in Q will be represented, as follows. From the root, there will be an outgoing edge labeled with the common k-name of Q, which will lead to a node. The subtree rooted at this node will have a leaf for each suffix having the core-name of Q as its prefix. It can be readily observed that each of these suffixes corresponds to some suffix (and a leaf) of $T(k)_1$. We will apply procedure CONTRACT below to $T(k)_0$ to include only these leaves and only those internal nodes which split into two (or more) paths in $T(k)_0$ leading from the root to two (or more) of these leaves.

Procedure $CONTRACT(T, s_1, s_2, \ldots, s_m)$
Input: A rooted tree $T(k)_0$. For each internal node of the tree, its children, whose number is denoted by n_{k+1}, are given in an array of size α. Assume that each node v has its distance from the root, denoted $level(v)$. We also assume that tree $T(k)_0$ has been preprocessed so that a query requesting the *lowest common ancestor* of any pair of nodes in $T(k)_0$ can be processed in constant time ([BV88], [SV88]). The input also includes $s_1, s_2, \ldots, s_m$, a subset of the leaves of T, given in the same order as they appear in T.

Output: a contracted version of $T(k)_0$ with the leaves $s_1, s_2, \ldots, s_m$.

3. **Get tree** $T(k)_2$. Consider the k-substrings whose k-cores were picked for $T(k)_1$ above. Consider the next k-substring. Pick the k-core that it spans, unless it is a suffix of some $(k+1)$-core. Given all the k-cores that were picked, we build the suffix tree of their suffixes using k-names and denote it $T(k)_2$. This is done using $T(k)_1$ similar to the construction of $T(k)_1$ using $T(k)_0$.

4. Use procedure MERGE to merge the three trees $T(k)_0, T(k)_1$, and $T(k)_2$ into $T(k)$, which is the suffix tree of the suffixes of all k-cores, using k-names. Procedure MERGE works in $4 \log^* n + 10$ rounds. Starting simultaneously from the roots of $T(k)_0, T(k)_1$, and $T(k)_2$ it advances step-by-step through the first $4 \log^* n + 10$ k-names in each of them, merging identical names into a single edge. By the first $4 \log^* n + 10$ k-names in each tree, we mean exploring in each tree all the paths starting from the root and in each path consider the first $4 \log^* n + 10$ k-names.

Perhaps the **most interesting observation in this paper** is that this gives the desired $T(k)$. This is implied by the following.

Observation. Take two k-suffixes that come from two different trees among $T(k)_0, T(k)_1$, and $T(k)_2$. Then, they can share a prefix of at most $4 \log^* n + 10$ k-cores.

Proof (for $k \geq 1$ only). This is implied by Lemma 1.2 as follows. Assume that the common prefix of two k-suffixes that come from two trees among $T(k)_0, T(k)_1$, and $T(k)_2$, is longer than $4 \log^* n + 10$ k-cores. Take the first such $4 \log^* n + 10$ k-cores, and focus on the chain of the $4 \log^* n + 10$ k-substrings that span them. They are identical in the two k-suffixes. By Lemma 1.2, the $(k+1)$-substrings that cover (well over $\log^* n$ of the) k-substrings in the middle section of this chain are also identical. Finally, consider the leftmost full $(k+1)$-cores in each of the k-suffixes. It is not hard to see that the $(k+1)$-suffix of each of these $(k+1)$-cores hits the two k-suffixes at the the same distance (which is $0, 1$, or 2) from their leftmost k-core. Therefore, both k-suffixes belong to the same tree among $T(k)_0, T(k)_1$, and $T(k)_2$. The proof for $k = 0$ is similar; actually, it becomes an important ingredient in proving the correctness of our whole algorithm.

References

[AILSV88] A. Apostolico, C. Iliopoulos, G. M. Landau, B. Schieber, and U. Vishkin, Parallel Construction of a Suffix Tree with Applications, In *Algorithmica*, 3: 347–365, 1988.

[BV88] O. Berkman, and U. Vishkin, Recursive star-tree parallel data-structure, In *SIAM J. Computing*, 22,2: 221–242, 1993.

[BDHPRS89] P. C. P. Bhatt, K. Diks, T. Hagerup, V. C. Prasad, T. Radzik, and S. Saxena, Improved Deterministic Parallel Integer Sorting, In *Information and Computation*, 94: 29–47, 1991.

[BLMPSZ91] G. E. Blelloch, C. E. Leiserson, B. M. Maggs, C. G. Plaxton, S. J Smith, M. Zagha, A Comparison of Sorting Algorithms for the Connection Machine CM-2, In *Proceedings of the 3rd Annual ACM Symposium on Parallel Algorithms and Architectures*, pages 3-16, 1991.

[CV86a] R. Cole, and U. Vishkin, Deterministic Coin Tossing with Applications to Parallel List Ranking, In *Information and Control*, 70: 32-53, 1986.

[CV86b] R. Cole, and U. Vishkin, Deterministic Coin Tossing and Accelerating Cascades: Micro and Macro Techniques for Designing Parallel Algorithms, In *Proceedings of the 18th Annual ACM Symposium on the Theory of Computing*, pages 206-219, 1986.

[Ga85] Z. Galil, Optimal Parallel Algorithms for String Matching, In *Information and Control*, 67: 144-157, 1985.

[KMR72] R. M. Karp, R. E. Miller, and A. L. Rosenberg, Rapid Identification of Repeated Patterns in Strings, Trees, and Arrays, In *Proceedings of the 4th Annual ACM Symposium on the Theory of Computing*, pages 125-136, 1972.

[MV91] Y. Matias, and U. Vishkin, On Parallel Hashing and Integer Sorting, In *Journal of Algorithms*, 12,4: 573-606, 1991.

[Mc76] E. M. McCreight, A Space - Economical Suffix Tree Construction Algorithm, In *Journal of the ACM*, 23: 262-272, 1976.

[MSU94] K. Mehlhorn, R. Sundar, and C. Uhrig, Maintaining Dynamic Sequences under Equality - Tests in Polylogarithmic Time, to appear In *Proceedings of the 5th Annual ACM - SIAM Symposium on Discrete Algorithms*, 1994.

[RR89] S. Rajasekaran, and J. H. Reif, Optimal and Sublogarithmic Time Randomized Parallel Sorting Algorithms, In *SIAM Journal of Computing*, 18: 594-607, 1989.

[SV93] S. C. Sahinalp, and U. Vishkin, Symmetry Breaking in Suffix Tree Construction, In preparation

[SV88] B. Schieber, and U. Vishkin, On Finding Lowest Common Ancestors: Simplification and Parallelization, In *SIAM Journal of Computing*, 17: 1253-1262, 1988.

[Vi85] U. Vishkin, Optimal Parallel Pattern Matching in Strings, In *Information and Control*, 67: 91-113, 1985.

[Vi91] U. Vishkin, Deterministic Sampling - A New Technique for Fast Pattern Matching, In *SIAM Journal of Computing*, 20: 22-40, 1991.

[We73] P. Weiner, Linear Pattern Matching Algorithm, In *Proceedings of the 14th IEEE Symposium on Switching and Automata Theory*, pages 1-11, 1973.

Appendix 1: Procedure LENGTH-BASED

We tentatively put a block divider to the (immediate) right of r_{j+k} and to the left of r_{j+1}, thereby separating the block partition of $r_{j+1} \ldots r_{j+k}$ from its surroundings. (Later, we may remove the tentative dividers put above if we find out that there is a divider to the left of r_j or to the right of r_{j+k+1}.)

We break into several cases depending on the remainder of k modula 4.

Case 1: k is an even number. Put block dividers to the right of each of $r_{j+2}, r_{j+4}, \ldots, r_{j+k}$.

Case 2: $k - 1$ is divisible by 4. The middle element of R_R is $r_{j+(k+1)/2}$. Put block dividers to the right of $r_{j+2}, r_{j+4}, \ldots, r_{j+(k-1)/2}$ and $r_{j+(k+5)/2}, r_{j+(k+9)/2}, \ldots, r_{j+k-2}$; so that the middle element is in a block of size 3 and all other blocks are of size 2.

Case 3: $k+1$ is divisible by 4. Put block dividers to the right of $r_{j+2}, r_{j+4}, \ldots, r_{j+(k-3)/2}$ and $r_{j+(k+3)/2}, r_{j+(k+7)/2}, \ldots, r_{j+k-2}$; so that again the middle element is in a block of size 3 and all other blocks are of size 2.

Procedure LENGTH-BASED actually continues to process R_R in iterations till it reduces to a single label. In each of the subsequent iterations the reduced length of R_R at the time will be treated according to the case analysis above; this is done in spite of the fact that a middle label may be different than other labels.

Example Let $R = \ldots 10000000002 \ldots$, the procedure LENGTH-BASED starts from the beginning and ending location of substring of 0s and partitions R accordingly (*Case 2* applies):

$\ldots 1|00|00|000|00|2 \ldots$

This ends procedure LENGTH-BASED.

SOME OPEN PROBLEMS IN APPROXIMATION

Mihalis Yannakakis

AT&T Bell Laboratories
Murray Hill, NJ 07974

Significant progress has been made in recent years in understanding better the limits of approximability for many important NP-hard combinatorial optimization problems. For any given problem we would like to know how close we can come to the optimal solution using a polynomial time approximation algorithm; closeness is usually measured by the worst-case ratio between the cost (or value) of the approximate solution and that of the optimal solution. A more modest goal is to classify problems qualitatively into one of the following three categories: (1) Those that can be approximated in polynomial time within *any* constant factor c of optimal, arbitrarily close to 1, i.e., problems that have a *polynomial time approximation scheme* (PTAS); (2) Those that can be approximated with *some* (but not every) constant ratio c; (3) Those that *cannot* be approximated within any constant ratio c, i.e., the best ratio grows with the input size n. As a result of developments in the last few years, we know now how to classify at least qualitatively some of the most basic and well-studied NP-hard problems, such as the Traveling Salesman Problem, Node Cover, Clique, Graph Coloring, Set Cover and many others.

Finding better approximation algorithms for natural, important problems will continue to be an active area of research for some time. We describe below some of our favorite open problems. We start with problems that have known constant factor approximation algorithms.

Traveling Salesman Problem (TSP) with triangle inequality. Given n cities and their pairwise distances, find the shortest tour. If the distances between the cities do not obey the triangle inequality but are arbitrary, then approximation within any factor is NP-hard [SG]. However, the more interesting case is when the triangle inequality is satisfied. In this case, a simple spanning tree heuristic achieves ratio 2. Christofides devised a more clever algorithm in 1976 which achieves ratio 3/2 [Ch]. No progress has been made since, and this remains still the best known guaranteed ratio. It should be possible to do better, at least in the Euclidean plane case.

There has been a lot of work by the Mathematical Programming community in finding good Linear Programming relaxations for the TSP (see eg. the book on the TSP edited by Lawler et al. [L+]). These relaxations provide lower bounds on the cost of the optimal tour. A bound that is good in practice and can be computed efficiently is the Held-Karp lower bound. It is known to be within 2/3 of the cost of the optimal tour, in fact of the tour computed by the Christofides heuristic [Wo, SW]; it is conjectured to come within 3/4 of the optimal cost, which may suggest the possibility of a 4/3 approximation algorithm. Finally, we note that the TSP with triangle inequality (even with all distances 1 or 2) is MAX SNP-hard [PY2] and hence does not have a polynomial time approximation scheme [A+].

Max Sat and *Max Cut*. In Max Sat we are given a set of clauses and want to find a truth assignment that satisfies the maximum number of clauses. If every clause has at least k literals, then we can achieve ratio $1 - 2^{-k}$ [J1]. In the general case the best known ratio is 3/4 [Y2, GW1]. In the case of Max 2Sat (2 literals per clause), Goemans and Williamson have managed very recently to improve the ratio to 0.87 using an elegant nonlinear programming relaxation [GW2]. They obtained the same ratio for the related Max Cut problem: given a graph, partition its nodes into two sets to maximize the number of edges in the cut. It is easy to achieve a ratio of 1/2 for Max Cut — many simple algorithms were proposed over the years with ratio 1/2, but nothing better was known until now. Max Sat and Max Cut are MAX SNP-hard [PY1] and thus cannot be approximated beyond a threshold $1 - \varepsilon$ (the current value of ε for Max 3Sat is around 1/65 [A+, BGLR, BSu]). In fact they are typical of problems in MAX SNP, and for this reason it is especially interesting to study methods for approximating them, which may then extend to other problems. It should be possible to improve on 3/4 for general Max Sat.

Node Cover. Given a graph, find the minimum set of nodes that cover all the edges. It is easy to achieve ratio 2. For example, the endpoints of a maximal matching give a 2-approximation. There are several other simple algorithms that achieve also ratio 2 even in the weighted case [BE, GP, Ho, Sa], but nothing better has been found.

The above problems are MAX SNP-hard, and thus we cannot hope for a ratio arbitrarily close to 1. As shown recently [KMSV], the class MAX SNP (closed under suitable L-reductions) coincides with the set APX of problems that have a constant factor approximation. The boundary between categories 1 and 2 is well captured by MAX SNP-hardness, in the sense that there is now a growing number and variety of MAX SNP-hard problems (eg. [BP, B+, D+, Kn]), so that one can reasonably expect to show that a problem does not have a polynomial time approximation scheme by reducing to it another known hard problem. Of course, this will not be always easy; even in the simpler case of just proving NP-completeness there are several problems that have been hard to resolve (eg, the Garey-Johnson list of open problems [GJ]).

The boundary between categories 2 and 3 is not yet as well and uniformly captured, and more work is required to refine and sharpen the tools. It is not clear that all problems in category 3, say Clique and Coloring on the one hand and Set Cover on the other, are nonapproximable for essentially the same reason. Their proof of nonapproximability uses probabilistically checkable proof systems. There are two types of such proof systems that are useful in reductions ([A+] on the one hand and [FL, LS] on the other, using respectively an arbitrary and a constant number of provers); for example, Clique and Graph

Coloring were shown nonapproximable using the first one and Set Cover using the second one [A+, LY1]. The two types are currently incomparable in strength. The latter type of proof system (based on a constant number of provers) is more structured and thus a more convenient starting point for reductions, but it gives currently weaker bounds on the degree of approximation and the evidence of nonapproximability is often weaker than P≠NP. As the technical results about these proof systems are being refined and improved, both the evidence and the degree of nonapproximability get strengthened [BGLR, FK, BSu]. It would be nice to have the least common denominator of [A+] and [FL]. In particular, a parameterized result for 2-prover, one-round systems roughly along the following lines (or some variant) may well be true.

Conjecture: There is a randomized polynomial time algorithm A which when given as input a 3SAT formula ϕ of size n and error parameter k, chooses randomly $O(\log n)$ bits, reads one entry each from two auxiliary "witness" tables T_1, T_2 and then accepts or rejects accordingly, so that the following property holds: There is a constant $c < 1$ such that for $2 \le k \le \Omega(\log n)$, if formula ϕ is satisfiable then there is a pair of tables T_1, T_2 of width k (i.e., every entry has k bits) such that A accepts with probability 1, whereas if ϕ is not satisfiable then for every pair of width-k tables, algorithm A accepts with probability at most c^k. Note that the claim for the $k=2$ end of the range is true - it is basically the result of [A+] for MAX SNP - and the $k = \Omega(\log n)$ end of the range would cover the [FL] result.

The following are some interesting problems that we do not know yet if they have a constant ratio approximation.

Asymmetric Traveling Salesman Problem (with triangle inequality). In the asymmetric TSP, the distance from city i to city j may be different than the distance from j to i. The best known approximation ratio is $\log n$ [FGM]. The Held-Karp lower bound in this case is also within the same $\log n$ factor of the optimal cost [Wi].

Separators, Sparse Cuts and Multicuts. There is a number of papers in the last few years that develop approximation algorithms for finding "small" cuts with nice properties, starting with [LR] for minimum cuts that divide the graph into two parts of roughly equal sizes; cuts that minimize the ratio of capacity over the demand across the cut [KARR, PT]; and minimum multicuts that separate given pairs of nodes [GVY]. The approximation ratio of these algorithms is $\log n$ or $\log^2 n$.

Node Deletion problems. Given a graph, we wish to find the minimum number of nodes whose deletion yields a subgraph that satisfies a desired property π. This problem is NP-hard for any nontrivial property π that is hereditary (preserved under node deletion), eg. planar, acyclic, bipartite etc. The complementary maximization problem, i.e. finding the maximum induced subgraph that has property π, is not approximable for all hereditary properties [LY2]. However, the minimization version is approximable within a constant factor for some properties π, while it appears not to be for others. For example, the node deletion problem corresponding to the property $\pi=$'independent set' is the Node Cover problem, which can be approximated with ratio 2. The node deletion problem corresponding to π='acyclic digraph' is the Feedback Node Set problem, for which the best known ratio is $O(\log^2 n)$ [LR]. It is not known whether the Feedback Node Set problem has a constant factor approximation. More generally, we would like to characterize the properties π for which the corresponding node deletion problem can be approximated within a constant factor. [LY2] contains a conjecture on where the dividing line is.

Graph Coloring for 3-colorable graphs, (or k-colorable for fixed k). For any ratio c, there is a k (which depends on c) such that we cannot have a c-approximate algorithm for k-colorable graphs [LY1]. However, we do not know whether the order of the two quantifiers in this statement can be switched; i.e. if there is a k such that no polynomial time algorithm can guarantee to color every k-colorable graph with $c \cdot k$ colors (for any c), and in particular if this holds for $k = 3$. A. Blum can color 3-colorable graphs with roughly $n^{3/8}$ colors [Bl], while [KLS] shows that it is NP-hard to do it with 5 colors.

For unrestricted number of colors we know that Graph Coloring cannot be approximated with ratio n^c for some c [LY1], and the same is true of the Clique problem [F+, AS, A+]; the current values for the exponent c are 1/4 for Clique and 1/10 for Coloring [BSu]. However, the upper bounds that are achieved by the known approximation algorithms are much weaker. It is conceivable that the lower bound results for Clique, Coloring (and other problems) could be sharpened to show that achieving ratio n^c is impossible for *every* $c < 1$ (instead of only some c), since no such algorithms are known so far. Bounds of this type are known to hold for some other graph problems, for example the node deletion problem for the property π='tree' and others [Y1, BS, Ir, LY2].

REFERENCES

[A+]   S. Arora, C. Lund, R. Motwani, M. Sudan, M. Szegedy, "Proof Verification and Hardness of Approximation Problems", Proc. 33rd IEEE Symp. on Foundations of Computer Science, 14-23, 1992.

[AS]   S. Arora, S. Safra, "Probabilistic Checking of Proofs", Proc. 33rd IEEE Symp. on Foundations of Computer Science, 2-13, 1992.

[AMP]   G. Ausiello, A. Marchetti-Spaccamela, M. Protasi, "Toward a Unified Approach for the Classification of NP-complete Optimization Problems", Theoretical Computer Science 12, 83-96, 1980.

[ADP]   G. Ausiello, A. D'Atri, M. Protasi, "Structure Preserving Reductions Among Convex Optimization Problems", J. Computer and System Sci. 21, 136-153, 1980.

[Ba]   L. Babai, "Transparent Proofs and Limits to Approximation", manuscript, 1993.

[BE]   R. Bar-Yehuda, S. Even, "A Linear Time Approximation Algorithm for the Weighted Vertex Cover Problem", J. Algorithms 2, 198-203, 1981.

[BGLR] M. Bellare, S. Goldwasser, C. Lund, A. Russel, "Efficient Probabilistically Checkable Proofs and Applications to Approximation", Proc. 25th ACM Symp. on Theory of Computing, 294-304, 1993.

[BR]   M. Bellare, P. Rogoway, "The Complexity of Approximating a Nonlinear Program", in Complexity in Numerical Optimization, P. Pardalos ed., World-Scientific, 1993.

[BSu]   M. Bellare, M. Sudan, "Improved Non-Approximability Results", draft, Nov. 1993.

[BS]   P. Berman, G. Schnitger, "On the Complexity of Approximating the Independent Set Problem", Information and Computation 96, 77-94, 1992.

[BP]   M. Bern, P. Plassman, "The Steiner Problem with Edge Lengths 1 and 2", Information Processing Letters, 32, 171-176, 1989.

[Bl]   A. Blum, "Some Tools for Approximate 3-coloring", Proc. 30th IEEE Symp. on Foundations of Computer Science, 554-562, 1990.

[B+]   A. Blum, T. Jiang, M. Li, J. Tromp, M. Yannakakis, "Linear Approximation of Shortest Superstrings", Proc. 23rd Annual ACM Symp. on Theory of Computing, 328-336, 1991.

[CP]   P. Crescenzi, A. Panconesi, "Completeness in Approximation Classes", Information and Computation 93, 241-262, 1991.

[Ch]   N. Christofides, "Worst-case Analysis of New Heuristics For the Traveling Salesman Problem", Technical Report, GSIA, Carnegie-Mellon, 1976.

[D+]   E. Dahlhaus, D. S. Johnson, C. H. Papadimitriou, P. Seymour, M. Yannakakis, "The Complexity of Multiway Cuts", Proc. 24th Annual ACM Symp. on Theory of Computing, 1992.

[FK]   U. Feige, J. Kilian, "Two Prover Protocols - Low Error at Affordable Rate", draft, Oct. 1993.

[FL]   U. Feige, L. Lovasz, "Two-prover One-round Proof Systems: Their Power and Their Problems", Proc. 24th Annual ACM Symp. on Theory of Computing, 733-744, 1992.

[F+]   U. Feige, S. Goldwasser, L. Lovasz, S. Safra, M. Szegedy, "Approximating Clique is Almost NP-complete", Proc. 32nd IEEE Symp. on Foundations of Computer Science, 2-12, 1991.

[FGM]   A. M. Frieze, G. Galbiati, F. Maffioli, "On the Worst-case Performance of Some Algorithms for the Asymmetric Traveling Salesman Problem", Networks 12, 23-39, 1982.

[GVY]   N. Garg, V. V. Vazirani, M. Yannakakis, "Approximate Max-Flow Min-(Multi)cut Theorems and Their Applications", Proc. 25th Annual ACM Symp. on Theory of Computing, pp. 698-707, 1992.

[GW1]   M. X. Goemans, D. P. Williamson, "A New 3/4 Approximation Algorithm for MAX SAT", Proc. 3rd Conf. on Integer Programming and Combinatorial Optimization, 1993.

[GW2]   M. X. Goemans, D. P. Williamson, ".878 Approximation Algorithms for MAX CUT and MAX 2SAT", draft, Nov. 1993.

[GJ]   M. R. Garey, D. S. Johnson, Computers and Intractability: A Guide to the Theory of NP-completeness, Freeman, 1979.

[GP]   D. Gusfield, L. Pitt, "Equivalent Approximation Algorithms for Node Cover", Information Processing Letters 22, 291-294, 1986.

[HK]   M. Held, R. M. Karp, "The Traveling Salesman Problem and Minimum Spanning Trees", Operations Research 18, 1138-1162, 1970.

[Ho]   D. S. Hochbaum, "Efficient Bounds for the Stable Set, Vertex Cover, and Set Packing Problems", Discrete Applied Mathematics 6, 243-254, 1982.

[Ir]   R. W. Irving, "On Approximating the Minimum Independent Dominating Set", Information Processing Letters 37, 197-200, 1991.

[J1]   D. S. Johnson, "Approximation Algorithms for Combinatorial Problems", J. Comp. Sys. Sc. 9, 256-278, 1974.

[J2]   D. S. Johnson, "Worst Case Behavior of Graph Coloring Algorithms", Proc. 5th Conf. on Combinatorics, Graph Theory and Computing, 513-527, 1974.

[J3]   D. S. Johnson, "The NP-completeness Column: An Ongoing Guide", J. of Algorithms 13, 502-524, 1992.

[Kn]   V. Kann, "On the Approximability of NP-complete Optimization Problems", Ph.D. Thesis, Royal Institute of Technology, Stockholm, 1992.

[Kr]   R. M. Karp, "Reducibility among Combinatorial Problems", in R. E. Miller and J. W. Thatcher (eds.), Complexity of Computer Computations, Plenum Press, 85-103, 1972.

[KARR] P. Klein, A. Agrawal, R. Ravi, S. Rao, "Approximation Through Multicommodity Flow", Proc. 31st Annual IEEE Symp. on Foundations of Computer Science, 726-737, 1990.

[KLS]   S. Khanna, N. Linial, S. Safra, "On the Hardness of Approximating the Chromatic Number", Proc. 2nd Israel Symp. on Theory and Computing Sys., 250-260, 1993.

[KMSV] S. Khanna, R. Motwani, M. Sudan, U. Vazirani, "On Syntactic versus Computational Views of Approximation", draft, Nov. 1993.

[KT]   P. G. Kolaitis, M,. N. Thakur, "Approximation Properties of NP Minimization Classes", Proc. 6th Conf. on Structures in Computer Science, 353-366, 1991.

[LS]   D. Lapidot, A. Shamir, "Fully Parallelized Multiprover Protocols for NEXPTIME", Proc. 32nd Annual IEEE Symp. on Foundations of Computer Science, 13-18, 1991.

[L+]   E. L. Lawler, J. K. Lenstra, A. H. G. Rinnoy Kan, D. B. Shmoys, The Traveling Salesman Problem, J. Wiley & Sons, 1985.

[LR]   F. T. Leighton, S. Rao, "An Approximate Max-flow Min-cut Theorem for Uniform Multicommodity Flow Problems with Applications to Approximation Algorithms", Proc. 28th Annual IEEE Symp. on Foundations of Computer Science, 256-269, 1988. Full (unpublished) version has additional results.

[LY1]   C. Lund, M. Yannakakis, "On the Hardness of Approximating Minimization Problems", Proc. 25th ACM Symp. on Theory of Computing, 286-293, 1993.

[LY2]   C. Lund, M. Yannakakis, "The Approximation of Maximum Subgraph Problems", Proc. 20th Intl. Coll. on Automata, Languages and Programming, 40-51, 1993.

[PR]   A. Panconesi, D. Ranjan, "Quantifiers and Approximation", Proc. 22nd ACM Symp. They of Computing, 446-456, 1990.

[Pa]   C. H. Papadimitriou, Computational Complexity, Addison-Wesley, 1993.

[PY1]   C. H. Papadimitriou, M. Yannakakis, "Optimization, Approximation and Complexity Classes", J. Computer and System Sci. 43, 425-440, 1991.

[PY2]   C. H. Papadimitriou, M. Yannakakis, "The Traveling Salesman Problem with Distances One and Two", Mathematics of Operations Research 18, 1-11, 1993.

[PM]   A. Paz, S. Moran, "Non Deterministic Polynomial Optimization Problems and their Approximation", Theoretical Computer Science 15, 251-277, 1981.

[PT]   S. Plotkin, E. Tardos, "Improved Bounds on the Max-Flow Min-Cut Ratio for Multicommodity Flow Problems", Proc. 25th Annual ACM Symp. on Theory of Computing, 691-697, 1993.

[SG]   S. Sahni, T. Gonzalez, "P-complete Approximation Problems", J. ACM 23, 555-565, 1976.

[Sa]   C. Savage, "Depth-first Search and the Vertex Cover Problem", Information Processing Letters 14, 233-235, 1982.

[SW]   D. B. Shmoys, D. P. Williamson, "Analyzing the Held-Karp TSP bound: A Monotonicity Property with Application", Information Processing Letters 35, 281-285, 1990.

[Wi]   D. P. Williamson, "Analysis of the Held-Karp Lower Bound for the Assymetric TSP", Operations Research Letters 12, 83-88, 1992.

[Wo]   L. A. Wolsey, "Heuristic Analysis, Linear Programming and Branch and Bound", Mathematical Programming Study 13, 121-134, 1980.

[Y1]   M. Yannakakis, "The Effect of a Connectivity Requirement on the Complexity of Maximum Subgraph Problems", J.ACM 26, 618-630, 1979.

[Y2]   M. Yannakakis, "On the Approximation of Maximum Satisfiability", Proc. 3rd ACM-SIAM Symp. on Discrete Algorithms, 1-9, 1992.

[Zu]   D. Zuckerman, "NP-complete Problems Have a Version That's Hard to Approximate", Proc. 8th Conf. on Structure in Complexity Theory, 305-312, 1993.

New Local Search Approximation Techniques for Maximum Generalized Satisfiability Problems [*]

Paola Alimonti

Dipartimento di Informatica e Sistemistica, Università di Roma "La Sapienza",
via Salaria 113, 00198 Roma, Italia

Abstract. We investigate the relationship beetween the classes MAX-NP and GLO by studying the Maximum Generalized Satisfiability problem, which is in the former class. We present a (2^{-B})-approximate greedy heuristic for this problem and show that no local search c-approximate algorithm exists, based on an h-bounded neighborhood and on the number of satisfied clauses as objective function. This implies that, with the standard definition of local search, MAX-NP is not contained in GLO.
We then show that, by introducing a different local search technique, that is using a different neighborhood structure for $B = 2$ and an auxiliary objective function in the general case, a local search (2^{-B})-approximate algorithms can be found for this problem. The latter result, that holds in the general case, suggests how to modify the definition of local search in order to extend the power of this general approch. In the same way, we can enlarge the class GLO of problems that can be efficiently approximated by local search techniques.

1 Introduction

It is well known that, although all NP-complete decision problems are polynomially isomorphic and in a certain sense "equally" difficult, the corresponding NP-hard optimization problems can behave in a very different way with respect to approximation properties. In fact, the polynomial time computable isomorphism between NP-complete problems is not strong enough to preserve the underlying structure of the optimization versions of these problem.

Namely, some problems can be approximated in polynomial time within any constant factor c of the optimum, that is they have a polynomial time approximation scheme (PTAS), others can be approximated within some constant c (APX), and some can not be approximated within any constant c unless $P = NP$ [3, 5, 11].

Among NP-complete problems, the Standard Satisfiability problem is the prototypical one. Moreover its optimization version, the Maximum Standard Satisfiability, plays an important role beetween NP-hard optimization problem [6, 7, 13]. In this paper we focus on the generalization of this problem, known as Generalized Satisfiability, and on its NP-hard optimization version, known as Maximum Generalized Satisfiability [12].

---
[*] Work supported by: the ESPRIT Basic Research Action No.7141 (ALCOM II); the Italian Project "Algoritmi, Modelli di Calcolo e Strutture Informative", Ministero dell'Università e della Ricerca Scientifica e Tecnologica; Consiglio Nazionale delle Ricerche, Italy.

The Generalized Satisfiability problem (G-SAT(B)) is expressed as follows:

Given a collection F of disjuntive form clauses, all of whose disjuncts are conjunctions containing up to B literals, determine whether or not there exists a truth assignment for F such that all clauses are simultaneously satisfied.

And the corresponding optimization problem, the Maximum Generalized Satisfiability problem (Max G-SAT(B)), is defined as follows:

Given a collection F of disjuntive form clauses, all of whose disjuncts are conjunctions containing up to B literals, find the truth assignment that satisfies the greatest number of clauses.

This problem is central in many areas of Computer Science (e.g. Computational Complexity, Logic and Artificial Intelligence); furthermore it seems to play a fundamental role in approximation theory.

Since the beginning of the 80's a remarkable amount of theoretical study has been carried out in the attempt of understanding the different approximation properties of NP optimization problems, and defining approximation classes. An early approach was mainly directed to explain the properties in combinatorial terms [1, 10]. Later, at the end of the 80's, a new approach, based on a logical point of view, was considered [12]. In this context two important classes were introduced based on Fagin's logical characterization of NP: MAX-NP and MAX-SNP. These classes include many NP optimization problems that are polynomially approximable within some constant c of the optimum, but have not polynomial time approximation schemas.

Recently, the combinatorial point of view has been again taken into consideration, and new classes have been defined. The class PLS (Polinomial-time Local Search), based on the complexity of finding the local optima, an the class GLO (Guaranted Local Optima), based on the quality of the local optima [2, 8]. For problems in the latter class the ratio between the value of any local optimum and the value of the global optimum is bounded by a constant.

While relationship beetween these different approximation classes has been studied [2], nothing was known on the relationship beetween GLO and MAX-NP.

The main contribution of this paper is to show that, using the standard definition of local search, MAX-NP is not contained in GLO. This result is obtained by considering the Maximum Generalized Satisfiability problem, which is in MAX-NP [12], and analyzing its behaviour with respect to approximation. More precisely, we first give an explicit greedy heuristic which allows us to determine a (2^{-B})-approximate solution for the problem. We then show that for Max G-SAT(B) there does not exist any local search heuristic for which 1) the objective function is the number of satisfied clauses, and 2) the neighborhood structure of a suitable solution is the set of solution obtained by changing a bounded number of variables. Therefore Max G-SAT(B) is not in GLO, and, hence, MAX-NP is not contained in GLO.

Finally we then consider the problem of approximating Max G-SAT(B) by a local search technique defined under two weaker sets of constrains (i.e. corresponding to classes larger than GLO). We first show that, by relaxing the definition of neighborhood, a local search heuristic exists for Max G-SAT(2). We then prove that, by relaxing the definition of local search (and, thus, using a different objective function) a general local search algorithm exists for Max G-SAT(B). The latter result open the question of whether MAX-NP is contained in such relaxed GLO class[1].

The reminder of the paper is organized as follows. In Section 2, we state the basic definitions and notations. In Section 3, we give an approximate greedy algorithm to solve Max G-SAT(B). In Section 4, we prove that Max G-SAT(B) is not in GLO. In Section 5, we relax the definition of neighborhood and present local search heuristic for $B = 2$. In Section 6, a weaker definition of local search is discussed, and a (2^{-B})-approximate local search algorithm for the general case is given. Finally, Section 7 contains conclusions and open problems.

2 Basic Definitions and Notation

In this section we formally define the Maximum Generalized Satisfiability problem [6], and report the main definitions from [2].

2.1 Maximum Generalized Satisfiability problem

Let $Y = \{y_1, y_2, ..., y_n\}$ be a set of Boolean variables. A *truth assignment* for Y is a function $t : Y \to \{True, False\}$. If $t(y) = True$, we say that y is true under t; if $t(y) = False$, we say that y is false under t. If y is a variable in Y, then y and $\bar{y}$ are literals over Y. The literal y is true under t if and only if the variable y is true under t; the literal $\bar{y}$ is true under t if and only if the variable y is false under t. We will denote by $T = \{p_1, ..., p_n\}$ the set of literals that are true under t. A *clause* c over Y is a disjunction of conjunctions of literals. It is satisfied by a truth assignment t if and only if all literals in at least one of its disjuncts are true under t. A *logical formula* $F = \{c_1, .., c_m\}$ is a collection of clauses and it is satisfiable if and only if there exists some truth assignment for Y that simultaneously satisfies all the clauses in F. Such a truth assignment is called a *satisfying truth assignment* for F.

The *Generalized Satisfiability* problem (G-SAT(B)) is defined as follows. Given a logical formula $F = \{c_1, ..., c_m\}$ of m disjuntive form clauses, formed by n Boolean variables $Y = \{y_1, y_2, ..., y_n\}$, all of whose disjuncts are conjunctions containing up to B literals, determine whether or not there exists a truth assignment for Y such that all clauses are simultaneously satisfied.

The *Maximum Generalized Satisfiability* problem (Max G-SAT(B)) is the optimization version of G-SAT(B). Given a logical formula $F = \{c_1, ..., c_m\}$ of m disjuntive form clauses, formed by n Boolean variables $Y = \{y_1, y_2, ..., y_n\}$, all of whose disjuncts are conjunction containing up to B literals, find the truth assignment for Y that satisfies the greatest number of clauses.

In the following of this work we consider the case in which each clause in the formula is composed by only one disjunct, that is by a conjunction containing up to B literals. Note that this does not affect the study of the worst case, as the presence of more disjuncts in a clause does not decrease the possibility that it is satisfied. Moreover every algorithm proposed can be easily extended to the general case.

---

[1] At the time of writing the final version of this paper we learned of the new result of Khanna et al. that show that also MAX-SNP is not contained in GLO, but any approximable NPO problem can be approximated via reduction by means of a generalized local search tecnique [9].

2.2 Approximation Classes and Local search

In the reminder of this section we report the main definitions related to NP optimization problems and local search algorithms [2].

Definition 2.1 *A NP optimization problem (NPO) P is a fourtuple (I, SOL, m, opt) such that:*

- *I is the set of instances of P and is recognizable in polynomial time;*
- *given an instance $x \in I$, $SOL(x)$ denotes the set of feasible solutions of x, there exists a polynomial q such that, given any $y \in SOL(x)$, $|y| \leq q(|x|)$;*
- *given an instance $x \in I$ and $y \in SOL(x)$, $m(x, y)$ is the measure of the feasible solution y of x;*
  m is computable in polynomial time;
  $opt \in \{min, max\}$.

$m^*(x)$ will denote the optimal value of an instance x.

In the following we consider only maximization problems; similar properties hold for minimization problems.

Definition 2.2 *Given an NP maximization problem $P = (I, SOL, m, max)$, an algorithm A is an approximate algorithm if for any given instance $x \in I$ it returns a feasible solution $A(x) \in SOL(x)$.*

Definition 2.3 *Given an NP maximization problem P and an approximate algorithm A, the performance ratio is*

$$R_A(x) = m(A(x))/m^*(x)$$

Definition 2.4 *Given an NP maximization problem P and an approximate algorithm A, A is said to be c-approximate (c constant, $0 < c < 1$) if its performance ratio $R_A(x)$ is such that $R_A(x) \leq c$.*

In this framework the following classes of NPO problems have been introduced [5]:

Definition 2.5 *Let P be an NPO problem.*

- *P belongs to the class APX if there exists a c-approximate algorithm A for P that runs in polynomial time.*
- *P belongs to the class PTAS if there exists an approximation algorithm A for P that takes an instance of the problem P and a value c as inputs and provides a c-approximate solution of x in time polynomial in the size of the instance x.*

Local search is an important techique used to compute polynomial time approximations for optimization for NP-hard optimization problem. Moreover relationships beetween structural and computational properties have been widely investigated, by considering the notion of local optima and their value with respect to global optima. In order to consider local search approximation algorithm we first report the following definitions.

Definition 2.6 *Given any NPO P, a neighborhood mapping is a function*

$$N : I \times SOL(I) \to \mathcal{P}(SOL(I))$$

such that

- *for any $x \in I$ and $y \in SOL(x)$, $N(x, y) \subseteq SOL(x)$ and $y \in N(x, y)$.*
- *there exists a polynomial q such that for every $x \in I$ and $y \in SOL(x)$, $|N(x, y)| \leq q(|x|)$. The set $N(x, y)$ is called neighborhood of y.*

Definition 2.7 *Let P be an NPO and let N be a neighborhood mapping; given an instance x and a feasible solution $y \in SOL(x)$, y is called a local optimum of x with respect to N if for every $z \in N(x, y)$ $m(x, z) \leq m(x, y)$. The value $m(x, y)$ is called local optimum value.*

Definition 2.8 *An NPO problem P has guaranteed local optima if there exists a neighborhood mapping N and a constant c such that, for every instance x, any local optimum y of x with respect to N has the property that $m(x, y) \geq c \cdot m^*(x)$.*

Definition 2.9 *An NPO problem P is polynomially bounded if there exists a polynomial r such that, given any instance $x \in I$, $m^*(x) \leq r(|x|)$.*

Definition 2.10 *Let P be an NPO. A neighbor y' of a feasible solution y is said to be a h-bounded neighbor if*

$$|(y - y') \cup (y' - y)| \leq h.$$

A neighborhood mapping N is said to be a h-bounded neighborhood mapping if there exists a constant h such that, given $x \in I$ and $y \in SOL(x)$, any neighbor $z \in N(x, y)$ is h-bounded.

Definition 2.11 *The class of all polynomially bounded NPO with guaranteed local optima with respect to h-bounded neighborhoods is called GLO.*

3 A Greedy Algorithm for Max G-SAT(B)

Since Max G-SAT(B) is in MAX-NP (and, thus, in APX), it follows that it can be approximated within some constant c [12]. In this section we present an explicit construction of a (2^{-B})-approximate greedy heuristic for the problem. This heuristic is a modification of an algorithm of Johnson for approximating Max SAT [7], and is partially derived from techniques in [12].

Intuitively, the 2^{-B} bound follows from the fact that, in the worst case, each of the m clauses of a formula F are composed of just one disjunct (i.e. each clause is a conjunction of at most B literals). In such case, the probability that each clause c_i ($1 \leq i \leq m$) is satisfied by a random truth assignment is $P_i = 2^{-|c_i|} \geq 2^{-B}$, which corresponds to the fraction of truth assignment that satisfies the clause. Then a random truth assignments satisfy on the average $m \cdot 2^{-B}$ clauses. Therefore, given a formula F, there exists a truth assignment that satisfies at least $m \cdot 2^{-B}$ clauses. A truth assignment that reaches such minimum guaranteed value can be determined by the following algorithm.

Algorithm 1

1. Assign to each clause $c \in F$ a weight $W(c) = 2^{-|c|}$. Set $SUB = \emptyset$, $TRUE = \emptyset$, $LIT = Y$, $LEFT = F$.

2. If no literal in any clause of $LEFT$ is in LIT, halt and return SUB.

3. Let y be a literal occurring in both LIT and a clause of $LEFT$. Let Y_T be the set of clause of $LEFT$ containing y, and Y_F be the set of clause of $LEFT$ containing $\bar{y}$

4. **If** $\sum_{c \in Y_T} W(c) \geq \sum_{c \in Y_F} W(c)$, **then**
       $TRUE \leftarrow TRUE \cup \{y\}$
       for each $c \in Y_T$ **then**
           $W(c) \leftarrow 2W(c)$
           if $W(c) = 1$ **then**
               $SUB \leftarrow SUB + 1$
               $LEFT \leftarrow LEFT - \{c\}$
       for each $c \in Y_F$ **then**
           $W(c) \leftarrow 0$
           $LEFT \leftarrow LEFT - \{c\}$
   else
       $TRUE \leftarrow TRUE \cup \{\bar{y}\}$
       for each $c \in Y_F$ **then**
           $W(c) \leftarrow 2W(c)$
           if $W(c) = 1$ **then**
               $SUB \leftarrow SUB + 1$
               $LEFT \leftarrow LEFT - \{c\}$
       for each $c \in Y_T$ **then**
           $W(c) \leftarrow 0$
           $LEFT \leftarrow LEFT - \{c\}$
   endif

5. Set $LIT = LIT - \{y, \bar{y}\}$. Go to 2.

The main idea underlying the previous algorithm is to associate a weight $W(c)$ to each clause c, corresponding to the probability that it is satisfied by a random truth assignment, that is $W(c) = 2^{-|c|}$. Then the choice of the value for a variable y is made considering the sums of the weights of the clauses containing the literal y, and of those containing the literal $\bar{y}$, respectively. This corresponds to making choices which, while the algorithm is executed, do not decrease the sum of the probabilities that each clause is satisfied by a random truth assignment.

Theorem 3.1 *The previous algorithm determines a* (2^{-B})*–approximate solution for Max G-SAT(B) in polynomial time.*

Proof. After the initialization step the probability that each clause $c \in S$ is satisfied by a random truth assignment is $W(c) = 2^{-|c|} \geq 2^{-B}$. Then, a random truth assignment satisfies on the average $W = \sum_{c \in S} W(c) \geq m \cdot 2^{-B}$ clauses. The algorithm proceedes by setting one variable at each step. Suppose that at a given step the variable y is set to true. Since in this case it must have been $\sum_{c \in Y_T} W(c) \geq \sum_{c \in Y_F} W(c)$, it follows immediatly that $W = \sum_{c \in S} W(c)$ can not decrease throughout the execution of the algorithm. Then, since at the end $W(c)$ is 1 for a satisfied clause, and 0 for an unsatisfied one, it follows immediately that the truth assignment satisfies at least $m \cdot 2^{-B}$ clauses. Hence, as the number of clauses $m \geq m^*$, the truth assignment satisfies at least $m^* \cdot 2^{-B}$ clauses.

As far as the running time of the algorithm is concerned, it is easy to see that, using binary search, the algorithm can be implemented in time $O(l \log l)$, where l is the size of the problem.

The previous algorithm can easily be extended to the general case by considering a weight for each disjunct in the clause and using the best one at each step.

Moreover, note that, as the choice of which literal to consider at step 3 is left largely undetermined, there is some potential for improvement which might affect its average case behaviour.

4 Local Search

Local search is one of the simplest general approaches that has met with empirical success to solve difficult combinatorial optimization problems. The general local search algorithm works starting from an initial feasible solution and trying to improve it searching for a better solution in its neighborhood. To analyze the behaviour of the Maximum Generalized Satisfiability problem with respect to local search we have first to define what h-neighborhood is in the case of Max G-SAT(B): given a set of boolean variables $Y = \{y_1, ..., y_n\}$, and a truth assignment T for Y, an h-bounded neighborhood for T is the set of truth assignment for Y such that at most the value of h variable is changed. In the following we prove that Max G-SAT(B) is not in GLO; that is, for any approximation ratio, whatever neighborhood size we allow, there are local optima with respect to such neighborhood that do not guarantee the ratio. In other words we show that for each $B > 1$, for any constant $h > 0$, and for any constant c, $0 \leq c \leq 1$, there exists an istance x of G-SAT(B) over a set of boolean variable Y, and a truth assignment T for Y, such that the number of clauses satisfied by T is $k < m^* \cdot c$, and T is a local optimum with respect to an h-bounded neighborhood.

To illustrate this consider the following examples for B=2.

1. Let be $F = \{(y_1 \wedge y_2), (y_1 \wedge y_3), (y_2 \wedge y_3), (\overline{y}_1 \wedge \overline{y}_4), (\overline{y}_1 \wedge \overline{y}_5), (\overline{y}_1 \wedge \overline{y}_6), (\overline{y}_1 \wedge \overline{y}_7), (\overline{y}_2 \wedge \overline{y}_8), (\overline{y}_2 \wedge \overline{y}_9), (\overline{y}_2 \wedge \overline{y}_{10}), (\overline{y}_2 \wedge \overline{y}_{11}), (\overline{y}_3 \wedge \overline{y}_{12}), (\overline{y}_3 \wedge \overline{y}_{13}), (\overline{y}_3 \wedge \overline{y}_{14}), (\overline{y}_3 \wedge \overline{y}_{15})\}$, and $T = \{y_i | 1 \leq i \leq 15\}$. The set of satisfied clauses S_0 is formed by the first 3 clauses, while the set of the unsatisfied clauses S_2 is formed by the other 12. The fraction of satisfied clauses is 1/5 and the number of variables to change to improve the solution is 4, that is T is a local optimum for $c = 1/5$ and $h = 3$. In fact to satisfy clauses in S_2 it needs to change the value of y_1 or y_2 or y_3, that is to make unsatisfied 2 clauses in S_0. Hence to improve the solution it needs to satisfy at least 3 clauses in S_2.
2. Let us now consider $F' = F \cup \{(y_1 \wedge y_a), (y_2 \wedge y_a), (y_3 \wedge y_a), (\overline{y}_1 \wedge \overline{y}_{16}), (\overline{y}_2 \wedge \overline{y}_{17}), (\overline{y}_3 \wedge \overline{y}_{18}), (\overline{y}_a \wedge \overline{y}_{a1}), (\overline{y}_a \wedge \overline{y}_{a2}), (\overline{y}_a \wedge \overline{y}_{a3}), (\overline{y}_a \wedge \overline{y}_{a4}), (\overline{y}_a \wedge \overline{y}_{a5})\}$ and $T' = T \cup \{y_a, y_{a1}, y_{16}, y_{17}, y_{18}\}$. The fraction of satisfied clauses is 3/13 and the number

of variables to change to improve the solution is 5, that is T' is a local optimum for $c = 3/13$ and $h = 4$.

The previous examples show that the number of variables which have to be changed to improve a solution and the fraction of unsatisfied clauses in a logical formula can be arbitrarily increased with the size of the problem. The following theorem formally proves such result.

Theorem 4.1 *For each $B > 1$, for any constant $h > 0$, and for any constant c, $0 \le c \le 1$, there exists an istance x of G-SAT(B) over a set of boolean variable Y, and a truth assignment T for Y, such that the number of clauses satisfied by T is $k < m^* \cdot c$, and T is a local optimum with respect to an h-bounded neighborhood.*

Proof. We will show that for each $B > 1$ and for any constant c, $0 \le c \le 1$, there exists an instance x of Max G-SAT(B), and a solution T, for which the number of satisfied clauses is $k < m^* \cdot c$ and T is a local optimum for any h-bounded neighborhood. We will prove that this statement holds for $k < m \cdot f(B)$. Then, as $m \cdot 2^{-B} \le m^*$, by letting $\cdot f(B) = 2^{-B} \cdot c$, we have $k < m \cdot 2^{-B} \cdot c \le m^* \cdot c$, which concludes the proof.

Consider now an istance of Max G-SAT(B), consisting in a set of m clauses $C = \{c_1, ..., c_m\}$ formed by a set of n variables. Let $T = \{p_1, ...p_n\}$, where $p_i = y_i$ or $p_i = \bar{y}_i$, be the initial solution, and k be the number of clauses satisfied by T. We construct the set of clauses C as follows: 1) the k clauses $c_1, ..., c_k$ are satisfied by T, and formed by all combination of order B of the first q literals $p_1, ..., p_q$, $k = \frac{q!}{B!(q-B)!}$; 2) the $m - k$ clauses $c_{k+1}, ..., c_m$ are formed by the literals $\bar{p}_1, ..., \bar{p}_n$ in the following way: in each clause there is one and only one literal $\bar{p}_i$, $1 \le i \le q$, and each literal $\bar{p}_j$, $q+1 \le j \le n$, appear in one and only one clause. Then $m - k = \frac{n-k}{B-1}$. Regardless of the value of $f(B)$, such construction implies $k < f(B) \cdot m$ for a suitable choice of $n > \frac{q!}{B!(q-B)!} \cdot [1 + \frac{(1-f(B))(B-1)}{f(B)}]$. Suppose now it is possible to increase the number of satisfied clauses by changing at most the truth value of h variables. We will show that h can increase with n for every value of B. To satisfy clauses in $\{c_{k+1}, ..., c_m\}$ we must modify at least one variable y_i, $1 \le i \le q$. But this implies to make unsatisfied $\frac{(q-1)!}{(B-1)!(q-B)!}$ clauses in $\{c_1, ..., c_k\}$. Moreover to satisfy $\frac{(q-1)!}{(B-1)!(q-B)!} + 1$ clauses in $\{c_{k+1}, ..., c_m\}$ we need to change $[\frac{(q-1)!}{(B-1)!(q-B)!} + 1] \cdot (B-1)$ variables y_j, $j > q$. Then $h > [\frac{(q-1)!}{(B-1)!(q-B)!} + 1] \cdot (B - 1)$, that can arbitrarialy increase with n.

As an immediate corollary of Theorem 4.1 we have the following results.

Corollary 4.2   a) *Max G-SAT(B) is not in GLO.*
      b) *MAX-NP is not contained in GLO.*

5   Relaxing the neighborhood constraint

In this section a local search heuristic is proposed to approximate Max G-SAT(2) based on a different neighborhood structure, which is bounded but not h-bounded. The neighborhood proposed allows us to determine a truth assignment that satisfies at least $\frac{m}{4}$ clauses in polynomial time. Consider a logical formula F of m clauses in G-SAT(2) over the set $Y = \{y_1, ..., y_n\}$ of boolean variable, and let T be a truth assignment for Y. Let S_0 be the set of satisfied clauses, S_1 be the set of unsatisfied clauses where one variable is true and one variable is false, and S_2 be the set of unsatisfied clauses where both variables are false.

The algorithm is based on the following lemma

Lemma 5.1 *Let F be a logical formula of m clauses in G-SAT(2) over the set $Y = \{y_1, ..., y_n\}$ of boolean variables, and let T be a truth assignment for Y. Then, if the number of satisfied clauses $K = |S_0| < \frac{m}{4}$, at least one of the following conditions holds: a) $2 \cdot |S_0| < |S_1|$; b) $|S_0| < |S_2|$.*

Proof. Let $K = |S_0| < \frac{m}{4}$, then $|S_1| + |S_2| > \frac{3 \cdot m}{4}$, that is $3 \cdot |S_0| < |S_1| + |S_2|$. Suppose now that $2 \cdot |S_0| \geq |S_1|$, and $|S_0| \geq |S_2|$. Therefore $3 \cdot |S_0| < |S_1| + |S_2| \leq 3 \cdot |S_0|$, a contradiction.

Now we superimpose a neighbohrood structure, and show that, if a solution for Max G-SAT(2) is a local optimum for this neighborhood, then it satisfies at least $K = |S_0| < \frac{m}{4}$ clauses.

Definition 5.1 *Let $Y = \{y_1, ..., y_n\}$ be a set of boolean variables, then given a truth assignment T for Y, the neighborhood N_T of T is the set of truth assignments for Y such that either the value of one and only one variable is changed or the value of every variable is changed; $N_T = \{(T'|(t'(y_i) = t(\overline{y_i}) \wedge t'(y_j) = t(y_j), 1 \leq j \leq n, j \neq i) \vee (t'(y_j) = t(\overline{y_j}), j = 1, ...n)\}$.*

Lemma 5.2 *Let F be a logical formula of m clauses in G-SAT(2) over the set $Y = \{y_1, ..., y_n\}$ of boolean variables, and T be an independently obtained truth assignment for Y. Then, if T is a local optimum for the neighborhood N_T, T satisfies at least $\frac{m}{4}$ clauses.*

Proof. Suppose $T = \{p_1, ...p_n\}$, where $p_i = y_i$ or $p_i = \overline{y}_i$, is a local optimum, and the number of satisfied clauses is $K = |S_0| < \frac{m}{4}$. From lemma 5.1 it must be either $2 \cdot |S_0| < |S_1|$, or $|S_0| < |S_2|$. If $2 \cdot |S_0| < |S_1|$, since $|S_0| = \sum_i \frac{|p_i|_{S_0}}{2}$, and $|S_1| = \sum_i |\overline{p}_i|_{S_1}$, we have $\sum_i |p_i|_{S_0} < \sum_i |\overline{p}_i|_{S_1}$, where $|p_i|_{S_j}$ ($|\overline{p}_i|_{S_j}$) is the cardinality of p_i ($\overline{p}_i$) in the set S_j. Then there exists a literals p_i such that $|p_i|_{S_0} < |\overline{p}_i|_{S_1}$, and changing the value of p_i the number of satisfied clauses increases. Now suppose $|S_0| < |S_2|$. Then changing the value of every literal the number of satisfied clauses increases. But by hypothesis T is a local optimum, a contradiction.

With this result we propose the following local search algorithm, which determines a truth assignment satisfying at least $\frac{m}{4}$ clauses starting from an independently obtained initial solution.

Algorithm 2

1. Let $TRUE$ be the set of true literals, and set $TRUE$ to an initial value. $TRUE = \{p_1, ..., p_n\}$, where $p_i = y_i$ or $p_i = \overline{y_i}$, for each $y_i \in Y$.
2. If $3 \cdot |S_0| \geq |S_1| + |S_2|$, halt and return $SUB = |S_0|$.
3. **If** $|S_1| > 2 \cdot |S_0|$, **then**

    $i \leftarrow 1$

    while $|p_i|_{S_0} \geq |\overline{p_i}|_{S_1}$ **do** $i \leftarrow i+1$

    $TRUE \leftarrow TRUE - \{p_i\} \cup \{\overline{p_i}\}$

    for each $c \in S_0$ **then**

        if $c = (p_i \wedge p_j)$ **then**

            $S_0 \leftarrow S_0 - \{c\}$

            $S_1 \leftarrow S_1 \cup \{c\}$

    for each $c \in S_1$ **then**

        if $c = (p_i \wedge \overline{p_j})$ **then**

            $S_1 \leftarrow S_1 - \{c\}$

            $S_2 \leftarrow S_2 \cup \{c\}$

        if $c = (\overline{p_i} \wedge p_j)$ **then**

            $S_1 \leftarrow S_1 - \{c\}$

            $S_0 \leftarrow S_0 \cup \{c\}$

    for each $c \in S_2$ **then**

        if $c = (\overline{p_i} \wedge \overline{p_j})$ **then**

            $S_2 \leftarrow S_2 - \{c\}$

            $S_1 \leftarrow S_1 \cup \{c\}$

else

    $TRUE \leftarrow \{\overline{p_1}, ..., \overline{p_n}\}$

    $AUX \leftarrow S_0$

    $S_0 \leftarrow S_2$

    $S_2 \leftarrow AUX$

endif

go to 2

Theorem 5.3 *The previous algorithm determines a (1/4)-approximate solution for Max G-SAT(2) in polynomial time.*

Proof. It follows immediately from Lemmata 5.1 and 5.2, and from $m \geq m^*$ that the algorithm is sound. Moreover it is easy to prove that it runs in polynomial time: as the number of satisfied clauses has to increase at each step, the number of step is no more than m, being the cost of each steps linear in the size of the formula.

6 Relaxing the local search constraint

In this section we show that, under a weaker definition of local search, it is possible to give a local search (2^{-B})-approximate heuristic for Max G-SAT(B). More precisely the algorithm we propose works using an auxiliary objective function, and then, in a certain sense, solving another problem, but without modifying the natural neighborhood structure.

Definition 6.1 *Let $Y = \{y_1, ..., y_n\}$ be a set of boolean variables, then given a truth assignment $T = \{p_1, ...p_n\}$ for Y, the neighborhood structure N_T of T is the set of truth assignments for Y such that the value of one and only one variable is changed;*
$N_T = \{(t(y_1), ..., t(y_n)|(t(y_i) = \overline{p_i} \wedge t(y_j) = p_j, 1 \leq j \leq n, j \neq i)\}$.

Given a truth assignment T for Y let S_i be the set of clauses where i literals are false in T and let $|p_l|s_i$ ($|\overline{p_i}|s_i$), $1 \leq l \leq n, 0 \leq i \leq B$, be the cardinality of the literal p_l ($\overline{p_i}$) is in S_i. Then, for $1 \leq i \leq B - 1$)

$$|S_0| = \sum_{l=1}^{n} \frac{|p_l|s_0}{B} \qquad |S_i| = \sum_{l=1}^{n} \frac{|p_l|s_i}{B-i} = \sum_{l=1}^{n} \frac{|\overline{p_i}|s_i}{i} \qquad |S_B| = \sum_{l=1}^{n} \frac{|\overline{p_i}|s_B}{B} \qquad (1)$$

Now, let $W = \sum_{i=0}^{B-1} L_i \cdot |S_i|$ be the weighted sum of the cardinalities of the S_i, where

$$L_i = \frac{1 + B \cdot L_{i+1} - (B - i - 1) \cdot L_{i+2}}{i+1} \quad (0 \leq i \leq B - 1) \qquad L_i = 0 \quad (i \geq B) \quad (2)$$

Then the following theorem holds:

Theorem 6.1 *Let F be a logical formula of m clauses in G-SAT(B) over the set $Y = \{y_1, ..., y_n\}$ of boolean variables, and $T = \{p_1, ..., p_n\}$, where $p_i = y_i$ or $p_i = \overline{y_i}$, be an indipendently obtained truth assignment for Y. If T is a local optimum for the neighborhood N_T and for the objective function W then T satisfies at least $m \cdot 2^{-B}$ clauses.*

Proof. Let ΔW be the variation of W and $\Delta|S_i|$ the variation of $|S_i|$ obtained by changing the value of one literal p_l in T. From 1, $\Delta W = \sum_{i=0}^{B} L_i \cdot \Delta|S_i| = L_0 \cdot (|\overline{p_i}|s_1 - |p_l|s_0) + \sum_{i=1}^{B-1} L_i \cdot (|p_l|s_{i-1} - |\overline{p_i}|s_i + |\overline{p_i}|s_{i+1} - |p_l|s_i) + L_B \cdot (|p_l|s_{B-1} - |\overline{p_i}|s_B)$.

Suppose now that it does not exist an index l such that changing the value of p_l $\Delta W > 0$, that is $\Delta W \leq 0$, for each l, $1 \leq l \leq n$. Then $L_0 \cdot (\sum_l |\overline{p_i}|s_1 - \sum_l |p_l|s_0) + \sum_{i=1}^{B-1} L_i \cdot (\sum_l |p_l|s_{i-1} - \sum_l |\overline{p_i}|s_i + \sum_l |\overline{p_i}|s_{i+1} - \sum_l |p_l|s_i) + L_B \cdot (\sum_l |p_l|s_{B-1} - \sum_l |\overline{p_i}|s_B) \leq 0$, and hence, from 1, $\sum_{i=0}^{B} A_i \cdot |S_i| \leq 0$, where $A_i = i \cdot L_{i-1} - B \cdot L_i + (B - i) \cdot L_{i+1}$ $0 \leq i \leq B$. Therefore, from 2, $A_i = 1$ $1 \leq i \leq B$, and $\sum_{i=1}^{B} |S_i| \leq (L_0 - L_1) \cdot B \cdot |S_0|$. Since, from 2, $L_i - L_{i+1} = \frac{1+(B-i-1)\cdot(L_{i+1}-L_{i+2})}{i+1}$, $0 \leq i \leq B - 1$, and $L_{B-1} - L_B = \frac{1}{B}$, we have $L_0 - L_1 = \sum_{k=1}^{B} \frac{(B-1)!}{k!(B-1-k)!}$, and $B \cdot (L_0 - L_1) + 1 = \sum_{k=0}^{B} \frac{B!}{k!(B-k)!} = 2^B$. Then $\sum_{i=1}^{B} S_i \leq (2^B - 1) \cdot |S_0|$.

Suppose now that $k = |S_0| < m \cdot 2^{-B}$. Then $\sum_{i=1}^{B} S_i > m \cdot \frac{(2^B-1)}{2^B}$ that is $\sum_{i=1}^{B} S_i > (2^B - 1) \cdot |S_0|$, a contradiction.

With this result we propose the following algorithm, which determines a truth assignment satisfying at least $m \cdot 2^{-B}$ clauses starting from an independently obtained initial solution.

Algorithm 3

1. Let $TRUE$ be the set of true literals, and set $TRUE$ to an initial value.
$TRUE = \{p_1, ..., p_n\}$, where $p_i = y_i$ or $p_i = \overline{y_i}$, for each $y_i \in Y$.
Let $\Delta W(l)$ be the W variation obtained by changing the value of one literal p_l in $TRUE$.
2. $i := 1;\ M = \emptyset$
   while $\Delta W(i) \leq 0$ **do** $i := i + 1$
   if $i = n + 1$ **then goto 4**
   $TRUE = TRUE - \{p_i\} \cup \{\overline{p_i}\}$
   for $j = 0$ **to** B **do**
       for each $c \in S_j$ **and** $c \notin M$ **then**
           if $p_i \in c$ **then**
               $M := M \cup \{c\}$
               $S_j := S_j - \{c\}$
               $S_{j+1} := S_{j+1} \cup \{c\}$
           if $\overline{p_i} \in c$ **then**
               $M := M \cup \{c\}$
               $S_j := S_j - \{c\}$
               $S_{j-1} := S_{j-1} \cup \{c\}$
3. **goto 2**
4. **return** $TRUE, S_0$
5. **halt**

The algorithm works in the following way: while it is possible to improve the function W by changing the value of one variable y_i, that is since there exists an index i, $1 \leq i \leq n$, for which $\Delta W(i) > 0$, it change the value of y_i and update the sets S_j, $0 \leq j \leq B$.

In order to prove that the previous algorithm runs in polinomial time, we have to show that the following lemma holds:

Lemma 6.2 $L_i > L_{i+1}$ for $i = 1, ..., B$

Proof. The lemma is proved by induction. From 1, we have $L_{B-1} = \frac{1 + B \cdot L_B}{B} > L_B$. Suppose now that $L_{i+1} > L_{i+2}$. Then $L_i = \frac{1 + B \cdot L_{i+1} - (B-i-1) \cdot L_{i+2}}{i+1} > \frac{1 + (i+1) \cdot L_{i+1}}{i+1} > L_{i+1}$

Theorem 6.3 *Algorithm 3 determines a (2^{-B})-approximate solution for Max G-SAT(B) in polynomial time.*

Proof. It follows immediately from Theorem 6.1 that the algorithm satisfies at least $m \cdot 2^{-B} \geq m^* \cdot 2^{-B}$ clauses. To show that it runs in polynomial time first observe that, as at each step we need to update every set S_i, $1 \leq i \leq B$, and each set contains at most m clauses composed by B literals, the cost of each step is bounded by $(m \cdot B \cdot (B+1))$. Moreover it must be $\Delta W > 0$ at least once every n steps, that is after considering all variables in Y, otherwise the algorithm stops. Consider now the function $\Delta V = B! \cdot \Delta W = B! \cdot \sum_{i=0}^{B} L_i \cdot \Delta |S_i|$ in which all coefficients are integers.

Therefore, since $\Delta W > 0$ and then $\Delta V > 0$, we have $\Delta V = 1$ at least once every n steps. Hence, as from Lemma 6.2 $V_{min} = B! \cdot L_B \cdot m$ (that is all the clauses are in the set S_B), and $V_{max} = B! \cdot L_0 \cdot m$ (that is all the clauses are in the set S_0), $\Delta V_{max} = V_{max} - V_{min} = B! \cdot L_0 \cdot m - (B-1)! \cdot m$, and then the maximum number of steps is bounded by $m \cdot (B-1)! \cdot (B \cdot L_0 - 1) \cdot n$.

7 Conclusions

The approximation properties of NP-hard optimizations problems have been widely studied, in the attempt of explaining their different behavior in terms of combinatorial or structural properties. In this framework different classes of NP-hard optimizations problems have been introduced.

In this work, in order to investigate the relationship beetween the classes MAX-NP and GLO, we have dealt with the Maximum Generalized Satisfiability problem, which is an important problem in the former class. First we have proposed a (2^{-B})-approximate greedy heuristic, and then we have considered its behaviour with respect to local search. In particular, we have shown that no local search c-approximate algorithm based on the number of satisfied clauses as objective function, and on an h-bounded neighborhood exists, that is the class MAX-NP is not contained in the class GLO. Howewer in the following it has been shown that considering a different neighborhood structure for $B = 2$, and using an auxiliary objective function in the general case, a local search (2^{-B})-approximate algorithms can be given for this problem. In particular, the latter result, that holds in the general case, suggest how to modify the definition of local search in order to extend the power of this general approch. In the same way we can enlarge the class GLO of problem that can be efficiently approximated by local search technique.

Acknowledgements: I would like to thank Giorgio Ausiello for his helpful comments and suggestions.

References

1. G. Ausiello, A. Marchetti-Spaccamela, M. Protasi, Toward a Unified Approach for the Classification of NP-Complete Optimization Problems, *Th. Comp. Sci.*,12,(1980), 83–96.
2. G.Ausiello, M. Protasi, NP Optimization Problems and Local Optima, *Technical Report*, Esprit Bra Alcom II, 1992.
3. D. Bruschi, D. Joseph, and P. Young, A Structural Overview of NP Opimization Problems, *Rapporto Interno n. 75/90*, Dip. di Scienze dell'Informazione, Univ. degli Studi di Milano, 1990.
4. S. A. Cook, The Complexity of Theorem Proving Procedures, *Proc. 3th. Annual ACM Symp. on Theory of Computing*, (1971), 151–158.
5. M. Garey, and D. Johnson, *Computers and Intractability: a Guide to the Theory of NP-Comleteness*, Freeman, San Francisco (1979).
6. P. Hansen, and B. Jaumard, Algorithms for the Maximum Satisfiability Problem, *Computing*, 44, (1990), 279–303

7. D. Johnson, Approssimation Algorithms for Combinatorial Problems, *J. Comp. Sys.*,Sc.9, (1974),256–278.

8. D.S. Johnson, C.H. Papadimitriou, M. Yannakakis, How Easy Is Local Search? *Journal of Computer and System Sciences*,37, (1988), 79–100.

9. S. Khanna, R. R. Motwani, M. Sudan, U. Vazirani, On Sintactic versus Computational Views of Approximability, Manuscript, 1993.

10. A. Paz, and S. Moran, Non Deterministic Polynomial Optimization Problems and their Approximation, *Th. Comp. Sci.*,15 (1981), 251–277.

11. C. Papadimitriou, and K. Steiglitz, *Combinatorial Optimization Algorithms and Optimization*, Prentice-Hall, Englewood Cliffs, New Jersey (1982).

12. C. Papadimitriou, and M. Yannakakis, Optimization, Approximation, and Complexity Classes, *Proc. 20th. Annual ACM Symp. on Theory of Computing*, (1988), 229–234. To appear J.Comp.Sys.Sc..

13. M. Yannakakis, On the Approximation of Maximum Satisfiability, *Proc. 3rd Annual ACM Symp. on Discrete Algorithm* ,(1992),1-9.

Learning Behaviors of Automata from Multiplicity and Equivalence Queries

F. Bergadano and S. Varricchio

Dipartimento di Matematica
Università di Catania
via A. Doria 6/A, 95100 Catania, Italy

Abstract

We consider the problem of identifying the behavior of an unknown automaton with multiplicity in the field Q of rational numbers (Q-automaton) from multiplicity and equivalence queries. We provide an algorithm which is polynomial in the size of the Q-automaton and of the maximum length of the given counterexamples. As a consequence, we have that Q-automata are PAC-learnable in polynomial time when multiplicity queries are allowed. A corollary of this result is that regular languages are polynomially predictable using membership queries w.r.t. the representation of unambiguous non-deterministic automata. This is important, as there are unambiguous automata such that the equivalent deterministic automaton has an exponentially larger number of states.

1 Introduction

Learning automata from examples and from queries has been extensively investigated in the past, and important results have been obtained recently [10,2,13,12]. Besides positive examples, an important input information comes from asking an oracle whether a particular string belongs to the target language. Such questions are called *membership queries*. Angluin [3] proves that, if we start from a set of strings that lead to every reachable state in the target automaton, a polynomial number of membership queries is sufficient for exact identification. However, if such a set of strings is not available, even if we know the size n of the target automaton, the needed number of queries is exponential in n.

Another possibility is found in *equivalence queries*: asking an oracle whether a guess is correct, and obtaining a counterexample if it is not. We shall also assume that the counterexamples have a maximum length m. It may be shown [5] that there is no polynomial time algorithm to exactly identify automata from equivalence queries only. However, there is a polynomial algorithm if both equivalence and membership queries are used [4].

Equivalence and membership queries then seem to be a necessary requirement for learning deterministic finite state automata. It remains to be seen if stronger formalisms may be learned under the same framework. This paper gives a positive answer in the direction of behaviors of non-deterministic finite state automata, i.e. functions assigning to every string the number of its accepting paths in a non-deterministic finite state acceptor. Such functions may be described with formal series, as defined in the next section, in the more general framework of automata with multiplicity.

2 Formal Series and Automata

Formal series and automata with multiplicity are the most important generalizations of the classical automata theory. In the last years their significant development helped in solving old problems in automata theory. In [11], using formal series, the decidability of the equivalence problem for deterministic multitape automata has been solved; in [16] a similar result has been shown for unambiguous regular languages in a free partially commutative monoid.

Let M be a non deterministic finite automaton, one can consider the so called *behavior* of M which is the map that associates to any word the number of its different accepting paths. More generally one can assign a multiplicity to the initial states, the final states and the edges of the automaton, so that the corresponding behavior must take into account the assigned multiplicities. In this way one constructs a theory which is general enough to contain, as particular objects, classical and probabilistic automata. In this context the theory of formal power series has been developed, giving a powerful algebraic tool for generalizing the concept of a formal language. Formal power series in non commuting variables have been extensively studied in Theoretical Computer Science and we refer to [8], [9], [14]; here we recall some notations and definitions.

Let K be a semiring and A^* the free monoid over a finite alphabet A; we consider the set $K\langle\langle A \rangle\rangle$ of all the applications $S : A^* \to K$. An element $S \in K\langle\langle A \rangle\rangle$ will also be denoted by $\sum_{u \in A^*} S(u)u$ or by $\sum_{u \in A^*} (S, u)u$ where (S, u) denotes $S(u)$. $K\langle\langle A \rangle\rangle$ is called the set of the *formal power series* over A^* with coefficients in K. In the sequel we shall consider formal series having coefficients in the rational field Q. We denote by $Q^{n \times n}$ the set of all square matrices $n \times n$ with entries in Q. We shall consider $Q^{n \times n}$ as having the structure of a monoid with respect to the row by column product; the identity matrix is denoted by Id. A map $\mu : A^* \to Q^{n \times n}$ is a monoid morphism if $\mu(\epsilon) = Id$ and for any $w \in A^+$, $w = a_1 a_2 \ldots a_n$, $a_i \in A$, one has $\mu(w) = \mu(a_1)\mu(a_2)\ldots\mu(a_n)$. A formal power series $S \in Q\langle\langle A \rangle\rangle$ is called *recognizable* or also *representable* if there exist a positive integer n, λ, $\gamma \in Q^n$ and a monoid morphism $\mu : A^* \to Q^{n \times n}$, such that for any $w \in A^*$

$$(S, w) = \lambda\mu(w)\gamma.$$

where λ and γ are to be considered row and column vectors, respectively. The triple (λ, μ, γ) is called a *linear representation* of S of dimension n, or also a Q-*automaton* for S.

A *non deterministic automaton* is a 5-tuple $M = (A, Q, E, I, F)$ where A is the input alphabet, Q is a finite set of *states*, $E \subseteq Q \times A \times Q$ is a set of *edges* and $I, F \subseteq Q$ are respectively the sets of the *initial* and *final* states. Let $w = a_1 a_2 \ldots a_n \in A^*$, an *accepting path* for w is any sequence $\pi = (p_1, a_1, p_2)(p_2, a_2, p_3) \ldots (p_n, a_n, p_{n+1})$, with $p_1 \in I$, $p_{n+1} \in F$ and $(p_i, a_i, p_{i+1}) \in$

E, for $1 \leq i \leq n$. The language accepted by M is the set $L(M)$ of all the words which have at least one accepting path; M is *unambiguous* if for any word $w \in L(M)$ there exists only one accepting path for w. We can associate to M a formal series $S_M \in \mathbb{Q}\langle\langle A\rangle\rangle$, also called the *behavior* of M, defined as follows: for any $w \in A^*$, (S_M, w) is the number of different paths which are accepting for w. Let $Q = 1, 2, \ldots, n$ and let $\lambda, \gamma \in \mathbb{Q}^n$ be the characteristic vectors respectively of I and F; consider the monoid morphism $\mu : A^* \to \mathbb{Q}^{n \times n}$, defined by $\mu(a)_{i,j} = 1$ if $(i, a, j) \in E$, and $\mu(a)_{i,j} = 0$ otherwise. Then one can easily prove (cf. [9]) that for any $w \in A^*$,

$$(S_M, w) = \lambda \mu(w) \gamma.$$

In particular S_M is representable and, when M is unambiguous, S_M corresponds to the characteristic function of $L(M)$. Finally, we shall need the following definitions:

Definition 1 *For any string $u \in A^*$, and a formal series S, the series S_u is defined by*
($\forall w \in A^*$) $(S_u, w) = (S, uw)$.

Definition 2 *For any set of strings $E \subseteq A^*$,*
$S \equiv_E T$ iff ($\forall w \in E$) $(S, w)=(T, w)$.
$S \equiv T$ stands for $S \equiv_{A^*} T$.

We shall use an oracle for answering *multiplicity queries* for any string w, i.e. for providing the value of (S, w), where S is the target formal series.

3 Observation Tables

Based on previous work by Angluin on deterministic finite state automata [4], we now introduce the concept of an observation table for a series S.

Definition 3 *Let $S \in \mathbb{Q}\langle\langle A\rangle\rangle$; an observation table is a triple*
$T = (P, E, T)$ where P and E are sets of strings, P is prefix-closed, E is suffix-closed, and $T:(P \cup PA)E \to \mathbb{Q}$ gives observed values of S, i.e., for all strings $w \in (P \cup PA)E$, $T(w)=(S, w)$.

Definition 4 *An observation table (P, E, T) is closed iff $\forall u \in P$, $\forall a \in A$, there are coefficients $\alpha_v \in \mathbb{Q}$, $v \in P$, such that*

$$S_{ua} \equiv_E \sum_{v \in P} \alpha_v S_v. \tag{1}$$

Definition 5 *An observation table (P, E, T) is consistent iff, for any choice of coefficients $\beta_v \in \mathbb{Q}$, $v \in P$,*

$$\sum_{v \in P} \beta_v S_v \equiv_E 0 \Rightarrow (\forall a \in A) \sum_{v \in P} \beta_v S_{va} \equiv_E 0. \tag{2}$$

Definition 6 *P is a complete set of strings for S iff* $\forall u \in P, \forall a \in A$, *there are coefficients* $\lambda_v \in \mathbb{Q}$, $v \in P$, *such that*

$$S_{ua} \equiv \sum_{v \in P} \lambda_v S_v. \tag{3}$$

When a table (P, E, T) is consistent and P is complete, the linear dependencies that are observed on E are valid for any string in A^*, as proved in the following:

Theorem 1 *Let (P,E,T) be a consistent observation table, where P is a complete set of strings for S, then*

$$\sum_{v \in P} \beta_v S_v \equiv_E 0 \Rightarrow \sum_{v \in P} \beta_v S_v \equiv 0. \tag{4}$$

As a consequence, the linear dependencies showing that the table is closed are also valid in A^*:

Corollary 1 *Let (P,E,T) be a consistent observation table, where P is a complete set of strings for S, then*

$$S_{ua} \equiv_E \sum_{v \in P} \lambda_v S_v \Rightarrow S_{ua} \equiv \sum_{v \in P} \lambda_v S_v. \tag{5}$$

4   The Learning Algorithm

The above results suggest that at any stage of the learning process when the table is closed and consistent, we may guess a series $M(P, E, T)$ by basing its representation upon the existing linear dependencies: .

- define $\{S^1, ..., S^k\} = \{S_u | u \in P\}$, with $S^1 = S_\epsilon = S$.

- for all $a \in A^*$, compute $\mu(a)$ satisfying

$$S_a^i \equiv_E \sum_j \mu(a)_{i,j} S^j. \tag{6}$$

Such a matrix exists because the table is closed.

- Construct the series M so as to make it depend from $\{S^1, ..., S^k\}$ with the same coefficients as $S=S^1$: let $\lambda = (1, 0, 0, ..., 0)$ and $\gamma_j = (S^j, \epsilon)$. The value of (S^j, ϵ) is found in the table since $S^j = S_u$ for some $u \in P$ and $\epsilon \in E$. Define the constructed series M with $(M, w) = \lambda \mu(w) \gamma$.

Theorem 2 *If P is a complete set of strings for S and (P,E,T) is a closed and consistent table, then $M(P, E, T) \equiv S$.*

We are now left with three problems: (1) closing a table (2) making a table consistent (3) making P complete. However, we will obtain completeness only indirectly and will return to it later.

4.1 Closing a table

Given a table (P, E, T), suppose S_{ua} is linearly independent from $\{S_v | v \in P\}$ with respect to E, in the sense that there are no coefficients $\lambda_{u,v}$ such that $S_{ua} \equiv_E \sum_{v \in P} \lambda_{u,v} S_v$. In this case ua is added to P, and the table is again checked for closure.

This procedure must terminate; more precisely, if the correct series S is representable with $(S, x) = \lambda \mu(x) \gamma$, where $\lambda, \gamma \in \mathcal{Q}^n$ and $\mu : A^* \to \mathcal{Q}^{n \times n}$ is a morphism, then at most n strings can be added to P when closing the table.

In fact, it should be noted that, when ua is added to P as indicated above, the dimension of $\{\lambda \mu(v) | v \in P\}$, as subset of the vectorial space $\mathcal{Q}^n$, is increased by one. Otherwise, $\lambda \mu(ua)$ would be equal to $\sum_{v \in P} \beta_v \lambda \mu(v)$ for some coefficients β_v and

$$(S_{ua}, x) = (S, uax) = \lambda \mu(ua) \mu(x) \gamma = \sum \beta_v \lambda \mu(v) \mu(x) \gamma = \sum \beta_v (S_v, x)$$

i.e., S_{ua} would depend linearly on $\{S_v | v \in P\}$. Since the dimension of $\{\lambda \mu(v) | v \in P\}$ is less than n, we cannot close the table more than n times. The above discussion does not depend on E.

4.2 Making tables consistent

Given a table (P, E, T) and a symbol $a \in A$, consider the two systems of linear equations:

$$(1) \sum_{v \in P} \beta_v S_v \equiv_E 0 \quad (2) \sum_{v \in P} \beta_v S_{va} \equiv_E 0.$$

with β_v as unknowns. Check if every solution of system (1) is also a solution of system (2). In this case the table is consistent. Otherwise, let β'_v, $v \in P$, be some solutions of (1) that are not solutions of (2) and $x \in E$ such that $\sum_{v \in P} \beta_v (S_{va}, x) \neq 0$. Add ax to E.

Suppose that S has a linear representation (λ, μ, γ) of dimension n; there cannot be more than n such additions to E, because every time a new string ax is added, the dimension of $\{\mu(w) \gamma | w \in E\}$ is increased by one. In fact, if $\mu(ax) \gamma = \sum_{w \in E} \delta_w \mu(w) \gamma$, then

$$\sum_{v \in P} \beta_v (S_{va}, x) = \sum_{v \in P} \beta_v \lambda \mu(v) \mu(ax) \gamma =$$

$$\sum_{v \in P} \beta_v \lambda \mu(v) \sum_{w \in E} \delta_w \mu(w) \gamma =$$

$$\sum_{w \in E} \delta_w \sum_{v \in P} \beta_v \lambda \mu(vw) \gamma =$$

$$\sum_{w \in E} \delta_w \sum_{v \in P} \beta_v (S_v, w) = 0$$

i.e., ax would not have been added to E.

4.3 The algorithm

We main now describe the procedure for exactly identifying S from multiplicity queries and counterexamples:

$\mathcal{T} \leftarrow (\{\epsilon\},\{\epsilon\},T)$, where $(T,\epsilon) = (S,\epsilon)$.
Repeat

- make the table closed and consistent (P and E are extended and the entries of T are filled in by multiplicity queries).

- ask for a counterexample t to $M(P,E,T)$ by means of an equivalence query.

- add t and its prefixes to P

until M is correct.

The main loop makes the table closed and consistent as described in the two previous subsections, and then constructs a guess M, which is based on the observed linear dependencies. We shall now prove that, if S has a linear representation (λ,μ,γ) of dimension n, after at most n equivalence queries, well have a correct guess, i.e. $M \equiv S$. We need the following:

Lemma 1 *Let $u \in P$ and $t \in uA^*$, and suppose that, for every x such that $t=uxz$, S_{ux} depends linearly from $\{S_v | v \in P\}$. Then, for every prefix ux of t,*

$$S_{ux} \equiv_E \sum_{v \in P} \mu(x)_{u,v} S_v. \tag{7}$$

where $\mu : A^* \rightarrow \mathbb{Q}^{k \times k}$ is the morphism corresponding to the observed linear dependencies as computed in (6), obtained when closing the table.

Theorem 3 *Let $(S,t) \neq (M,t)$. Then there is a prefix t_0 of t such that S_{t_0} is linearly independent from $\{S_v | v \in P\}$.*

Corollary 2 *If S has a linear representation of dimension n, then after at most n iterations, the algorithm stops*

4.4 Complexity analysis

All we need to do is reorganize some of the previous results and determine the complexity of computing the linear dependencies. Let n be the dimension of a linear representation of S, one has:

- The main loop in the algorithm is repeated at most n times (corollary 2).

- $|E| \leq n$ (section 4.2).

- For the cardinality of P, the discussion is slightly more involved. From the discussions of sections 4.1 and 4.3, we see that every time either (1) a string is added to P while closing the table or (2) a counterexample is processed, the dimension of $\{\lambda\mu(v)|v \in P\}$ is increased by at least one. The worst case is when this always happens with the counterexamples and the main loop in the algorithm is repeated exactly n times, because also the prefixes of the counterexamples need to be added to P. If m is the maximum length of a counterexample, then, $|P| \leq nm$.

- For every $a \in A$, the table needs to be closed. This amounts to solving a system of $|E|$ equations in $|P|$ unknowns; in the worst case, nm simultaneous equations in nm unknowns. This is an operation of complexity $O(n^3m^3)$ [1]. If $|A|=k$, the complexity of closing the table is $O(kn^3m^3)$.

- Checking for consistency was described in section 4.2, and requires the algorithm to confront the two systems of equations

$$(1) \sum_{v \in P} \beta_v S_v \equiv_E 0 \quad (2) \sum_{v \in P} \beta_v S_{va} \equiv_E 0.$$

with β_v as unknowns. The table is consistent if every solution of (1) is also a solution of (2). This is the same as checking whether the $|E|$ equations of system (2) do not add additional constraints, i.e. if the corresponding vectors of coefficients depend linearly from the ones of system (1). This operation, in the worst case, requires solving n systems of nm equations in nm unknowns, with complexity $O(n^4m^3)$.

Therefore the complexity of the algorithm is $O(kn^4m^3 + n^5m^3)$. This, together with the fact that, when the main loop terminates, $M \equiv S$, establishes our main result:

Theorem 4 *Representable series may be exactly identified in polynomial time from queries and counterexamples.*

This may be seen as a generalization of Angluin's result for finite state automata [4]. It should be noted that theorem 2 was the inspiring idea behind the algorithm, but has not been used to obtain the above result. In particular, completeness of P was not directly verified in the algorithm nor used in the proofs. However, when the algorithm terminates, the set P must be complete:

Theorem 5 *When $M \equiv S$,*

$$S_{ua} \equiv \sum_{v \in P} \mu(a)_{u,v} S_v \tag{8}$$

As a consequence, when the guess M is correct, P must be complete.

Therefore completeness of P may be seen as a characterization of success when learning S. Moreover, it may be noted that, if we start with a set P which is already complete, by virtue of theorem 2, S may be exactly identified in polynomial time by means of multiplicity queries only. The algorithm is as follows:

$T \leftarrow (P, \{\epsilon\}, T)$, where T is filled in with multiplicity queries.
make the table consistent, and fill missing values with queries
output $M(T)$.

The table will certainly be closed, as P is complete, and, by theorem 2, the
final guess M must be correct. This result should be confronted with [3], where
a similar framework is described for finite state automata: a regular language
may be exactly identified in polynomial time, from membership queries only, if
we are given a complete set of representants for Nerode's equivalence classes.

5   PAC-learnability

If we are not interested in the exact identification of Q-automata, but only
in a probably approximately correct (PAC) series, equivalence queries are not
necessary, as they can be substituted by sampling of example strings with their
correct multiplicity. The technique is analogous to the one used for DFAs in [4].
As a consequence, Q-automata are PAC-learnable, when multiplicity queries are
allowed.

If a non-deterministic automaton is unambiguous, the corresponding formal series S will be such that, for any string w, (S, w) is wither 0 or 1. In this
case, then, multiplicity queries reduce to ordinary membership queries. Suppose now that a regular language L is recognized by an unambiguous NFA M,
with a corresponding formal series S. The present paper gives an algorithm
for PAC-learning a representation of S in polynomial time w.r.t. the number
of states of M, if membership queries are allowed. In other words, following the usual terminology (see, e.g., [12]), regular languages are polynomially
predictable using membership queries w.r.t. the representation of unambiguous non-deterministic automata. The importance of this lies in the fact that
there are unambiguous NFAs such that the equivalent DFA has an exponentially larger number of states [15]. We then have a substantial improvement
over previous results establishing predictability w.r.t. to a deterministic representation.

References

[1] A. V. Aho, J. E. Hopcroft and J. D. Ullman, *The Design and Analysis of Computer Algorithms* Addison-Wesley, 1974.

[2] D. Angluin. On the complexity of minimum inference of regular sets. *Information and Control*, 39:337–350, 1978.

[3] D. Angluin. A note on the number of queries needed to identify regular languages. *Information and Control*, 51:76–87, 1981.

[4] D. Angluin. Learning regular sets from queries and counterexamples. *Information and Computation*, 75:87–106, 1987.

[5] D. Angluin. Negative results for equivalence queries. *Machine Learning*, 5:121–150, 1990.

[6] F. Bergadano and A. Giordana and L. Saitta. *Machine Learning: an Integrated Framework and its Applications* Ellis Horwood, 1991.

[7] F. Bergadano and D. Gunetti. *An Interactive System to Learn Functional Logic Programs. Proc. Int. Joint Conf. on Artificial Intelligence*, Morgan Kaufmann, 1993.

[8] J. Berstel and C. Reutenauer. *Rational series and their languages* Springer-Verlag, Berlin, 1988.

[9] S. Eilenberg *Automata, Languages and Machines* Vol. A, Academic Press, New York, 1974.

[10] M. E. Gold. Complexity of automaton identification from given data. *Information and Control*, 37:302–320, 1978.

[11] T. Harju and J. Karhumaki. Decidability of the multiplicity equivalence problem of multitape finite automata *Proc. 22nd STOC*, 477-481, (1990).

[12] B. K. Natarajan. *Machine Learning: a Theoretical Approach* Morgan Kaufmann, 1991.

[13] L. Pitt and M. K. Warmuth. The Minimum Consistent DFA Problem Cannot be Approximated within any Polynomial. *Journal of the ACM*, 40:95-142, 1993.

[14] A. Salomaa and M. Soittola. *Automata theoretic aspects of formal power series* Springer-Verlag, New York, 1978.

[15] R. E. Stearns and H. B. Hunt. On the Equivalence and Containment Problems for Unambiguous Regular Expressions, Regular Grammars and Finite Automata *SIAM J. Comput.*, 14:3,598-611, 1985.

[16] S. Varricchio. On the decidability of the equivalence problem for partially commutative rational power series *Theoretical Computer Science*, 99: 291-299, 1992.

Measures of Boolean Function Complexity
Based on Harmonic Analysis

A. Bernasconi[1] and B. Codenotti[1]

Istituto di Elaborazione dell'Informazione,
Consiglio Nazionale delle Ricerche,
56100 - Pisa (Italy)

Abstract. In this paper we study some measures of complexity for Boolean functions based on Abstract Harmonic Analysis. More precisely, we extend the notion of average sensitivity of Boolean functions and introduce the generalized sensitivity (or *k-sensitivity*) which accounts for the changes in the function value when *k* input bits are flipped. We find the connection between *k*-sensitivity and Fourier coefficients. We then analyze some properties of the 2-sensitivity and in particular its relation with the average sensitivity (or 1-sensitivity).

1 Introduction

Over the last decade, methods from harmonic analysis on the Boolean cube have been used to prove some general results on the *complexity of Boolean functions*. In fact the Fourier transform of a Boolean function, which is a linear map of the vector of function values onto a set of rational *Fourier coefficients*, gives information about the *regularities* of the function, and thus also about its *computational complexity*.

An important feature of the Fourier coefficients is their relation with the *sensitivity*, which is a useful tool for a computational complexity investigation. The sensitivity measures if the value of a function is likely to change in correspondence to arguments which are Hamming neighbors (see [KKL 88], [LMN 89], [BOH 90], [BC 93]).

In this paper we continue the investigation of sensitivity, which we have started in a previous paper [BC 93]. More precisely, we extend the notion of average sensitivity of Boolean functions and introduce the generalized sensitivity (or *k-sensitivity*) which accounts for the changes in the function value when exactly *k* input bits are flipped.

We now give some background on abstract harmonic analysis and we review the notion of sensitivity. For a more comprehensive treatment, see [Leh 71], [Loo 53] and [BC 93].

Let us consider the space $\mathcal{F}$ of all the two-valued functions on $\{0, 1\}^n$. The domain of $\mathcal{F}$ is a locally compact Abelian group and the elements of its range, i.e. 0 and 1, can be added and multiplied as complex numbers. The above properties allow one to analyze $\mathcal{F}$ by using tools from harmonic analysis. This means that it is possible to construct an orthogonal basis set of Fourier transform kernel functions for $\mathcal{F}$. An orthogonal basis is provided by the set of functions $\{Q_w | w \in \{0, 1\}^n\}$, where $Q_w(x) = (-1)^{w_1 x_1}(-1)^{w_2 x_2} \ldots (-1)^{w_n x_n} = (-1)^{w^T x}$. These functions are known as *group characters* or Fourier transform kernel functions.

We can now define the *Abstract Fourier Transform* of a Boolean function f as the rational valued function f^* which defines the coordinates of f with respect to the basis $\{Q_w(x), w \in \{0, 1\}^n\}$, i.e., $f^*(w) = 2^{-n} \sum_x Q_w(x) f(x)$. Then $f(x) = \sum_w Q_w(x) f^*(w)$ is the Fourier expansion of f.

It is possible to derive a convenient matrix formulation for the transform pair. Let us consider a $2^n \times 2^n$ matrix H_n whose (i, j)-th entry h_{ij} satisfies $h_{ij} = (-1)^{i^T j}$, where $i^T j$ denotes the inner

---

[*] This work has been partly supported by the Italian National Research Council, under the "Progetto Finalizzato Sistemi Informatici e Calcolo Parallelo", subproject 2 "Processori dedicati".

product of the binary expansions of i and j. If we use the binary 2^n-tuple representation for f and f^*, i.e. $f = [f_0 \, f_1 \ldots f_{2^n-1}]^T$ and $f^* = [f_0^* \, f_1^* \ldots f_{2^n-1}^*]^T$, then, from the fact that $H_n^{-1} = 2^{-n} H_n$, we get $f = H_n f^*$ and $f^* = 2^{-n} H_n f$.

Note that the matrix H_n is the Hadamard symmetric transform matrix [Leh 71] and can be recursively defined as

$$H_1 = \begin{pmatrix} 1 & 1 \\ 1 & -1 \end{pmatrix} , \qquad H_n = \begin{pmatrix} H_{n-1} & H_{n-1} \\ H_{n-1} & -H_{n-1} \end{pmatrix} .$$

The Fourier coefficients have the following interpretation. The coefficient f_0^* is the probability that f takes the value 1. The first order coefficients f_j^*, with $j = 2^{n-i}$, measure the correlation of the function f with its i-th variable. In general, the coefficients f_j^* measure the correlation of the function with the parity of those variables whose position corresponds to a 1 in the binary expansion of j. There is no correlation if $f_j^* = 0$ and maximum correlation if $|f_j^*| = \frac{1}{2}$. The sign of the coefficient indicates if the correlation is actually with the parity ($f_j^* = -\frac{1}{2}$) or with its complement ($f_j^* = \frac{1}{2}$).

We can now recall the notions of *influence* and *sensitivity*.

Definition 1 ([KKL 88].) *Let $\mathcal{A}$ be a set of variables. The* influence *of $\mathcal{A}$ on f, denoted by $I_f(\mathcal{A})$, is the probability that f remains undetermined as long as the variables in $\mathcal{A}$ are not assigned values and the other variables are assigned at random according to the uniform distribution.*

Definition 2 ([LMN 89].) *The* sensitivity $s_w(f)$ *of a Boolean function f on a string $w \in \{0, 1\}^n$ is the number of Hamming neighbors $\hat{w}$ of w such that $f(w) \neq f(\hat{w})$. The* average sensitivity *of f, $s(f)$, is the average of $s_w(f)$ over all $w \in \{0, 1\}^n$. $s(f)$ can also be defined as the sum of the influences of all the variables on f.*

We also recall the notion of *maximal sensitivity*, $s_{max}(f)$, which is defined as the maximum of $s_w(f)$ over all $w \in \{0, 1\}^n$.

The Fourier analysis can be used to determine how much a function is sensitive to its arguments. More precisely, in terms of the Fourier transform of f, the average sensitivity becomes

$$s(f) = 4 \sum_w |w| \, (f^*(w))^2 ,$$

as shown in [HMM 82], [KKL 88] and in [BC 93].

We now review some results related to our work.

Karpovsky [Kar 76] proposes to use the number of nonvanishing Fourier coefficients of a Boolean function f as a measure of its complexity. Hurst et al. [HMM 82] relate the circuit complexity of a Boolean function to its power spectrum coefficients. Brandman et al. [BOH 90] establish a relationship between the Fourier coefficients of a Boolean function f and (i) the average size of any decision tree for f; (ii) the minimum number of $\wedge$ gates in a circuit computing f according to its disjuntive normal form. Kahan et al. [KKL 88] find connections between influence and harmonic analysis and use theorems on influence to prove results on rapidly mixing Markov chains.

Linial et al. [LMN 89] take advantage of the relation between the average sensitivity of Boolean functions and their Fourier transform to prove several facts. Their main result is that an AC^0 Boolean function has almost all of its *power spectrum* on the low-order coefficients. In fact, they prove that the sum of squares associated with all high Fourier coefficients is very small. This result implies several new properties of functions in AC^0. One of these properties is that every function f in AC^0 has low average sensitivity, i.e. $s(f) = O((\log n)^{d+3})$, where d is the depth of the circuit.

Applications of the Fourier representation to computational learning are shown in [LMN 89] and in [KM 91].

Bruck [Bru 90] and Bruck and Smolensky [BS 92] use abstract harmonic analysis to derive necessary and sufficient conditions for a Boolean function to be a polynomial threshold.

Wegener [Weg 87] introduces the notion of *critical complexity*, which coincides with our notion of maximal sensitivity, and Simon [Sim 83] shows that every function that depends on all its variables has maximal sensitivity of at least $\Omega(\log n)$.

With respect to the application of sensitivity to *parallel complexity*, in [CDR 86] and [Nis 89], the CREW PRAM complexity of Boolean functions computations is expressed in terms of sensitivity.

In [BC 93] it is analyzed the distribution of the average sensitivity over the set of all the n-ary Boolean functions and it is proposed to apply sensitivity to the study of special reductions in the class NC^1. Moreover, it is proved that sets in AC^0 whose characteristic function is symmetric are almost sparse or co-sparse, and have exponentially decreasing average sensitivity.

In this paper we explore some notions derived from the Fourier analysis of Boolean functions. The goal is to provide a tool which potentially can lead to complexity results.

We first extend the notion of average sensitivity of Boolean functions and introduce the generalized sensitivity (or *k-sensitivity*). We find the connection between k-sensitivity and Fourier coefficients (see Lemma 1 in Section 2). We then show that the k-sensitivity, normalized in a suitable way, turns out to be a *submodular measure of complexity*, according to the definition of *formal complexity measure* as introduced in [Raz 92] and in § 8.8 in [Weg 87] (see Lemma 2 and 3 in Section 2). We then analyze some properties of the 2-sensitivity and in particular its relation with the average sensitivity. We show that the 2-sensitivity behaves as a sort of derivative of the average sensitivity (see Lemma 4 in Section 2). We find an upper bound for the sum of the average sensitivity and the 2-sensitivity for functions in the class $AC^0 \cup pAC^0$, defined in [BC 93] (see Theorem 1 in Section 2). We also analyse the value of 2-sensitivity for symmetric Boolean functions.

Finally we start an investigation of the connections between the Fourier coefficients and the complexity of Boolean functions. We study some properties of random Boolean functions by providing an estimate of the expected value and the variance of their Fourier coefficients (see Lemma 6 in Section 4). Furthermore, we show which properties of the coefficients are likely to express the fact that a function is, e.g., computable by polynomial size circuits (see Theorem 4 in Section 4).

We will use the following notations. Unless otherwise specified, the indexing of vectors and matrices starts from 0 rather than 1. All the logarithms are to the base 2. The notation *polylogn* stands for a function growing like a polynomial in the logarithm of n. Given a Boolean function f on k binary variables, we will often use its 2^k-tuple vector representation $f = (f_0 \, f_1 \ldots f_{2^k-1})$, where $f_i = f(x(i))$ and $x(i)$ is the binary expansion of i. If x and y are two binary strings of the same length, then $d(x, y)$ and $x \oplus y$ denote their Hamming distance and the string obtained by computing the *exclusive or* of the bits of x and y, respectively. $|x|$ denotes the number of ones in a binary string x. We say that two strings are k-neighbors if their Hamming distance is k.

The rest of this paper is organized as follows. In section 2 we introduce the definition of *k-sensitivity* and we find its relation with the Fourier coefficients. We then prove that the k-sensitivity, normalized in a suitable way, is a *submodular measure of complexity*. Finally, we prove some results related to the 2-sensitivity. In section 3 we consider the special case of symmetric functions. In section 4 we use the information contained in the Fourier coefficients of Boolean functions to prove some complexity results. In section 5 we provide a framework for future research.

2 Generalized Sensitivity of Boolean Functions and Fourier Coefficients

In this section we introduce the notion of generalized sensitivity (or *k-sensitivity*) and we find its relation with the Fourier coefficients. We then prove that the k-sensitivity, normalized in a suitable way, is a *submodular measure of complexity*. Finally, we prove some results on the 2-sensitivity.

The average sensitivity is a measure of how a function behaves when one input bit is flipped. We consider here a generalization of the average sensitivity by allowing a few - rather than just one - bits to be flipped. This may be useful in order to distinguish functions with high average sensitivity but with an overall very regular behaviour, like PARITY, from functions which are hard to compute.

Let f be a Boolean function defined on $\{0, 1\}^n$.

Definition 3 *The k-sensitivity $s_w^{(k)}(f)$ of f on a string $w \in \{0, 1\}^n$ is the number of strings $\hat{w}$ such that $d(w, \hat{w}) = k$ and $f(w) \neq f(\hat{w})$, i.e.*

$$s_w^{(k)}(f) = \sum_{|u|=k} (f(w) - f(w \oplus u))^2.$$

The k-sensitivity of f, $s^{(k)}(f)$, is the average of $s_w^{(k)}(f)$ over all $w \in \{0, 1\}^n$:

$$s^{(k)}(f) = \frac{1}{2^n} \sum_w s_w^{(k)}(f).$$

Finally, the maximal k-sensitivity, $s_{max}^{(k)}(f)$, is the maximum of $s_w^{(k)}(f)$ over all $w \in \{0, 1\}^n$.

Note that the 1-sensitivity coincides with the average sensitivity. Moreover, as for the average sensitivity, the k-sensitivity can assume only rational values.

Lemma 1. *Let f be a Boolean function defined on $\{0, 1\}^n$, and let f^* be the vector of its Fourier coefficients. Then*

$$s^{(k)}(f) = 4 \sum_w W_k(|w|) (f^*(w))^2,$$

where the weight function $W_k(|w|)$ is defined as follows

$$W_k(|w|) = \sum_{l=0}^{\lceil \frac{k}{2} \rceil - 1} \binom{|w|}{2l+1} \binom{n-|w|}{k-(2l+1)}.$$

Proof. We take advantage of two identities, i.e. *Parseval's identity:* $\sum_v f(v) = 2^n \sum_v (f^*(v))^2$, and *autocorrelation identity:* $\sum_v f(v)f(v \oplus u) = 2^n \sum_v Q_v(u) (f^*(v))^2$, which is a consequence of the orthogonality of the functions $Q_v(u)$.

We obtain, for the k-sensitivity $s_w^{(k)}(f)$ of a string w,

$$s_w^{(k)}(f) = \sum_{|u|=k} (f(w) - f(w \oplus u))^2 = \sum_{|u|=k} (f(w) + f(w \oplus u) - 2f(w)f(w \oplus u)),$$

from which

$$s_w^{(k)}(f) = \binom{n}{k} f(w) + \sum_{|u|=k} f(w \oplus u) - 2 \sum_{|u|=k} f(w)f(w \oplus u).$$

For the k-sensitivity of the function f, some algebra yields

$$s^{(k)}(f) = 2^{-n+1} \left[\binom{n}{k} \sum_w f(w) - \sum_{|u|=k} \sum_w f(w)f(w \oplus u) \right].$$

Since we have

$$\sum_{|u|=k} \sum_w f(w)f(w \oplus u) = 2^n \binom{n}{k} \sum_w (f^*(w))^2 - 2^{n+1} \sum_w W_k(|w|) (f^*(w))^2,$$

then, by Parseval's equality, we obtain

$$s^{(k)}(f) = 4 \sum_w W_k(|w|) (f^*(w))^2.$$

The explicit expression of the k-sensitivity in terms of the Fourier coefficients, for $1 \leq k \leq 4$, is given by:

$$s^{(1)}(f) = 4 \sum_w |w| \, (f^*(w))^2$$

$$s^{(2)}(f) = 4 \sum_w |w| \, (n - |w|) \, (f^*(w))^2$$

$$s^{(3)}(f) = 4 \sum_w \left[|w| \binom{n - |w|}{2} + \binom{|w|}{3} \right] (f^*(w))^2$$

$$s^{(4)}(f) = 4 \sum_w \left[|w| \binom{n - |w|}{3} + (n - |w|) \binom{|w|}{3} \right] (f^*(w))^2.$$

We now show that the k-sensitivity, normalized in a suitable way, turns out to be a *submodular formal measure of complexity* (see [Raz 92] and § 8.8 in [Weg 87]).

First of all, we recall the definition of *formal complexity measure* and of *submodular formal complexity measure*.

Definition 4 ([Raz 92].) *A nonnegative real-valued function μ defined on the set of all Boolean functions in n variables is a* formal complexity measure *if*

$$\mu(x_i) \leq 1, \qquad \mu(\neg x_i) \leq 1, \qquad (1 \leq i \leq n). \tag{1}$$

$$\mu(f \vee g) \leq \mu(f) + \mu(g) \qquad \text{for each } f, \, g; \tag{2}$$

$$\mu(f \wedge g) \leq \mu(f) + \mu(g) \qquad \text{for each } f, \, g. \tag{3}$$

Definition 5 ([Raz 92].) *A nonnegative real-valued function μ defined on the set of all Boolean functions in n variables is a* submodular formal complexity measure *if it satisfies the condition* (1) *and the* submodularity condition

$$\mu(f \vee g) + \mu(f \wedge g) \leq \mu(f) + \mu(g) \qquad \text{for each } f, \, g \tag{4}$$

which is stronger than both (2) *and* (3).

Lemma 2. *The average sensitivity, $s(f)$, is a submodular measure of complexity.*

Proof. Conditions (1) and (4) are easily seen to be satisfied if one considers the definition of the average sensitivity of a Boolean function as sum of the influences of its variables.

Let us consider the functions $f_i(x_1, x_2, \ldots, x_n) = x_i$, for $1 \leq i \leq n$. Since for $1 \leq i, j \leq n$ $I_{f_i}(x_j) = \delta_{ij}$, we get $s(f_i) = 1$ and condition (1) is satisfied.

Now, let f and g be two Boolean functions defined on $\{0, 1\}^n$. Let us consider the functions $(f \vee g)(x_1, x_2, \ldots, x_n)$ and $(f \wedge g)(x_1, x_2, \ldots, x_n)$. By using some probabilistic arguments related to the definition of influence, we obtain

$$I_{f \vee g}(x_i) + I_{f \wedge g}(x_i) = I_f(x_i) + I_g(x_i), \qquad 1 \leq i \leq n.$$

Then, from the fact that the average sensitivity can be defined as sum of the influences of the variables, we get

$$s(f \vee g) + s(f \wedge g) = s(f) + s(g),$$

and condition (4) is satisfied.

With respect to the k-sensitivity, for $2 \leq k \leq n$, one can easily verify that, since the k-sensitivity of the function $f(x_1, x_2, \ldots, x_n) = x_i$ is equal to the binomial coefficient $\binom{n-1}{k-1}$, then condition (1) is not satisfied. In order to obtain a measure of complexity, we can normalize $s^{(k)}(f)$ as follows:

$$\hat{s}^{(k)}(f) = \binom{n-1}{k-1}^{-1} s^{(k)}(f).$$

Lemma 3. *The normalized k-sensitivity, $\hat{s}^{(k)}(f)$, is a submodular measure of complexity.*

Proof. Condition (1) is obviously satisfied. Let f and g be two Boolean functions defined on $\{0, 1\}^n$. For all $w \in \{0, 1\}^n$ we get

$$\hat{s}_w^{(k)}(f) + \hat{s}_w^{(k)}(g) = \sum_{|u|=k} [(f(w) - f(w \oplus u))^2 + (g(w) - g(w \oplus u))^2].$$

Now, let us consider the functions $(f \vee g)(w) = f(w) \vee g(w)$ and $(f \wedge g)(w) = f(w) \wedge g(w)$. For all $w \in \{0, 1\}^n$ we have

$$\hat{s}_w^{(k)}(f \vee g) = \sum_{|u|=k} (f(w) \vee g(w) - f(w \oplus u) \vee g(w \oplus u))^2;$$

and

$$\hat{s}_w^{(k)}(f \wedge g) = \sum_{|u|=k} (f(w) \wedge g(w) - f(w \oplus u) \wedge g(w \oplus u))^2.$$

From these expressions is easy to verify, e.g. by considering all the sixteen possible combinations of the value of $f(w)$, $g(w)$, $f(w \oplus u)$ and $g(w \oplus u)$, that

$$\hat{s}_w^{(k)}(f \wedge g) + \hat{s}_w^{(k)}(f \vee g) \leq \hat{s}_w^{(k)}(f) + \hat{s}_w^{(k)}(g).$$

Then, by averaging over all w, the submodularity condition is easily seen to be satisfied.

From the above proof, it follows that the normalized maximal k-sensitivity is a submodular formal measure of complexity.

To combine the information carried by the k-sensitivities, one can also consider their normalized sum

$$\hat{S}(f) = \frac{1}{n} \sum_{k=1}^{n} \binom{n-1}{k-1}^{-1} s^{(k)}(f),$$

which is again a submodular measure of complexity.

We now focus our attention on the 2-sensitivity.

First of all, note that the maximum value achievable by $s^{(2)}(f)$ does not coincide with the maximum value achievable by $s_{max}^{(2)}(f)$, i.e. $\binom{n}{2}$. This follows from neighborhood properties of strings in the n-cube, i.e. intersections between strings at Hamming distance 2. In general these intersections take place whenever we consider even Hamming distances. For this reason, for even k, we have $0 \leq s^{(k)}(f) < \binom{n}{k}$. For odd k, we have $0 \leq s^{(k)}(f) \leq \binom{n}{k}$ and the maximum value $\binom{n}{k}$ is achieved by the PARITY function and its complement.

With respect to the 2-sensitivity, we have that $\frac{n^2}{4} \leq \max_{f \in \mathcal{F}}\{s^{(2)}(f)\} < \binom{n}{2}$, where the value $\frac{n^2}{4}$ is achieved by the parity function computed on $\frac{n}{2}$ input bits (for simplicity, we have assumed that n is even; when n is odd, we can consider the parity function computed on $\frac{n-1}{2}$ input bits, which has 2-sensitivity equal to $\frac{n^2-1}{4}$). Moreover, the distribution of the 2-sensitivity of all Boolean functions of n variables is characterized by an expected value equal to $\frac{1}{2}\binom{n}{2}$ and by a variance equal to $\frac{1}{2^{n+2}}\binom{n}{2}$.

The next Lemma shows that the 2-sensitivity behaves as a sort of derivative of the average sensitivity.

Lemma 4. *Let f be a Boolean function defined on $\{0,1\}^n$. Then*

$$\frac{1}{2^n} \sum_w \sum_{i=1}^n \left| \frac{ds_w^{(1)}(f)}{di} \right| \leq 2 \ s^{(2)}(f),$$

where $\frac{d}{di}(s_w^{(1)}(f)) = s_w^{(1)}(f) - s_{w \oplus i}^{(1)}(f)$, and $w \oplus i$ denotes the string obtained from w by flipping the i-th bit.

Proof. Follows from the definition of 2-sensitivity of a function f on a string $w \in \{0,1\}^n$ and by using some algebra.

We now present an upper bound on the sum of the average sensitivity and the 2-sensitivity for the functions in the class sAC^0 (*symmetric AC^0*), where

$$sAC^0 = AC^0 \cup pAC^0 \quad \text{and} \quad pAC^0 = \{g \mid g = f \oplus PARITY, \ f \in AC^0\}$$

(see [BC 93]).

Theorem 1 *For any function $f \in sAC^0$ we have $[s^{(1)} + s^{(2)}](f) \leq n \ \log^{O(1)} n$.*

Proof. For any function $f \in AC^0$ we have

$$[s^{(1)} + s^{(2)}](f) = (n+1) \ s^{(1)}(f) - 4 \sum_w |w| \ (f^*(w))^2 \ \leq \ n \ s^{(1)}(f).$$

Then the thesis follows from the fact that Boolean functions in AC^0 have average sensitivity $s^{(1)}(f) \leq \log^{O(1)} n$ [LMN 89].
Let us now consider a function g in pAC^0. Then $g = f \oplus PARITY$, with $f \in AC^0$, and $s^{(1)}(g) = n - s^{(1)}(f)$ [BC 93]. Moreover, from the properties of the Fourier transform of functions in pAC^0, we obtain $s^{(2)}(g) = s^{(2)}(f)$, and the thesis easily follows.

Note that this result removes the differences in the behaviour of the sensitivity of functions in AC^0 and pAC^0.

3 Applications to Symmetric Functions

In this section we consider the special case of symmetric Boolean functions. We first point out the different behaviour of the k-sensitivity for k even and odd. We then show that $s^{(2)}(f)$ cannot achieve its maximum value over the set of the symmetric Boolean functions. Finally, we present two alternative expressions for $s^{(2)}(f)$, one of which related to the *spectrum* of f (see [FKPS 85] and [BC 93]).

Let w be a binary string of length n, and let $\hat{w}$ denote its k-neighbors. For even $k = 2h$, there are $\binom{n-|w|}{h} \binom{|w|}{h}$ k-neighbors $\hat{w}$ such that $|\hat{w}| = |w|$. Therefore, for any symmetric Boolean function f, we have $s^{(2h)}(f) \leq \binom{n}{2h} - \frac{1}{2^n} \sum_w \binom{n-|w|}{h} \binom{|w|}{h}$. On the contrary, for odd k, there exist symmetric Boolean functions with k-sensitivity of the order of $\binom{n}{k}$, e.g. the parity function.

By using this result, we can prove the following Lemma.

Lemma 5. *The 2-sensitivity $s^{(2)}(f)$ can not achieve its maximum value over the set of the symmetric Boolean functions.*

Proof. First of all, we have that for any symmetric function

$$s^{(2)}(f) \leq \binom{n}{2} - \frac{1}{2^n} \sum_w (n - |w|) \ |w| = \frac{1}{2} \binom{n}{2}.$$

Then, the thesis follows from the existence of non symmetric Boolean functions with 2-sensitivity greater than $\frac{1}{2}\binom{n}{2}$, e.g. the parity function computed on $\lfloor \frac{n}{2} \rfloor$ input bits.

Note that the bound given by the previous Lemma is tight, in fact the value $\frac{1}{2}\binom{n}{2}$ can be detected by looking at functions with spectra of the form 00110011...0011.

As for the average sensitivity (see [BC 93]), the 2-sensitivity of symmetric functions can be evaluated in a simplified way. We get

$$s^{(2)}(f) = 4n(n-1) \sum_{i=1}^{n} \binom{n-2}{i-1} (f^*_{2^i-1})^2,$$

to be compared with a similar expression for the average sensitivity, i.e.

$$s(f) = 4n \sum_{i=1}^{n} \binom{n-1}{i-1} (f^*_{2^i-1})^2.$$

In terms of the spectrum of f, we get

$$s^{(2)}(f) = \frac{1}{2^n} \sum_{i=0}^{n} \binom{n}{i} \left\{ \binom{i}{2} |w_i - w_{i-2}| + \binom{n-i}{2} |w_i - w_{i+2}| \right\},$$

where w_i denotes the i-th character of the spectrum of f. Starting from this expression, it is possible to find the relation existing between the complexity measure for symmetric functions introduced in [FKPS 85] and the 2-sensitivity.

4 Fourier Coefficients and the Complexity of Boolean Functions

We first show some properties of the Fourier coefficients of random Boolean functions and then present some computational consequences of the *Quotient Group Character Theorem* (see [Leh 71]).

We say that a Boolean function $f : \{0, 1\}^n \rightarrow \{0, 1\}$ is a random Boolean function if, for all $w \in \{0, 1\}^n$, we have $\Pr\{f(w) = 0\} = \Pr\{f(w) = 1\} = \frac{1}{2}$.

The next Lemma gives the characterization of the Fourier coefficients of a random function.

Lemma 6. *Let f be a random Boolean function. Then*

(a) *The expected value of each, but the first, Fourier coefficient is equal to 0.*
(b) *The variance of each Fourier coefficient is equal to $\frac{1}{2^{n+2}}$.*

Proof. (a) Note that $E[f(w)] = \frac{1}{2}$. Then, the thesis easily follows from the matrix formulation of the Fourier transform $f^* = 2^{-n} H_n f$, by exploiting the structural properties of the Hadamard symmetric transform matrix H_n and the linearity of the expected value. With respect to the first Fourier coefficient, which represents the probability that f takes the value 1, we obviously obtain $E[f^*_0] = \frac{1}{2}$.

(b) From the definition of variance, and from the fact that $E[f^*_j] = \frac{1}{2}\delta_{0j}$, for $0 \leq j \leq 2^n - 1$, we get $V[f^*_j] = E[(f^*_j)^2] - \frac{1}{4}\delta_{0j}$. Then, the thesis follows by evaluating $E[(f^*_j)^2]$. We can work as in part (a), by exploiting the properties of H_n and the linearity of the expected value.

Note that, from the value of the variance and from the fact that it must result $\sum_{j=0}^{2^n-1}(f^*_j)^2 \sim \frac{1}{2}$, it follows that the Fourier coefficients of a random function are indeed "small", but they vanish with low probability. More precisely, one can verify, by looking at the structure of the Hadamard matrix, that the probability for a coefficient to be zero is of the order of $\binom{2^{n-1}}{2^{n-2}}^2 \binom{2^n}{2^{n-1}}^{-1} \sim \frac{1}{\sqrt{\pi 2^{n-3}}}$. Note also that these results agree with the interpretation of the Fourier coefficients as measures of the correlation between the function and the parity of subsets of the input bits (see Section 1).

We now present some computational consequences of the *Quotient Group Character Theorem* [Leh 71].

Theorem 2 *(Quotient Group Characters [Leh 71], pp.142-143.) Let $\mathcal{F}$ be the set of all Boolean functions defined on $\{0,1\}^n$. Let $V \subseteq \{0,1\}^n$. Then the subset of Fourier basis functions $\{Q_w(x)|x \in V\}$ is a complete orthogonal basis for the subspace of $\mathcal{F}$ consisting of all functions which are constant on cosets of V.*

Corollary 3 *([Leh 71], page 144.) Let $f \in \{0,1\}^n$. Assume that the set $\{w|f^*(w) \neq 0\}$ has rank $n - k$. Then f can be defined as a function of at most $n - k$ variables, each of them being a linear combination of the n original variables.*

Theorem 4 *Let $f = \{f_i, i = 1, 2, ...\}$ be a family of Boolean functions, where f_i is defined on $\{0,1\}^i$. If the cardinality of the set of nonzero Fourier coefficients of f_i is at most $i^{O(1)}$, then the set associated to f_i belongs to the class P/poly, i.e. non-uniform P. If the cardinality of the set of nonzero Fourier coefficients of f_i is upper bounded by a constant, then f can be computed by circuits of logarithmic depth.*

Proof. In the first case, the thesis follows from the fact that $f_i(w)$, for any i and any $w \in \{0,1\}^n$, can be evaluated by the inner product between one row of the Hadamard matrix H_i and the vector of nonzero Fourier coefficients. Note that the nonzero Fourier coefficients can be inserted in a non-uniform circuit, since their number is at most polynomial.

In the second case, the thesis follows from Corollary 9. For each i, f_i can be defined as function of a constant number of variables. Then, once computed the linear combinations of the i original variables by a circuit of logarithmic depth, each f_i can be computed by a circuit of depth independent on i.

5   Concluding Remarks

In this paper we have continued the investigation of relations between circuit complexity and Fourier analysis of Boolean Functions. The starting point of this investigation has been the observation that the notion of average sensitivity seems to be too weak to allow complexity results for Boolean functions, besides, e.g., the fact that functions with high average sensitivity are not in AC^0. We have thus introduced more general notions, which give a more detailed information on the function. We have found the relation between these notions and the Fourier coefficients.

Future work to be done includes the attempt of using these measures to find complexity results. For example, we would like to characterize the i-sensitivity; $i = 1, 2, \ldots, k$, of functions in complexity classes between AC^0 and NC^1.

References

[BC 93]   A. Bernasconi, B. Codenotti. *Sensitivity of Boolean Functions, Abstract Harmonic Analysis, and Circuit Complexity.* Technical Report TR-93-030 International Computer Science Institute, Berkeley, CA. (1993).

[BOH 90]   Y. Brandman, A. Orlitsky, J. Hennessy. *A Spectral Lower Bound Technique for the Size of Decision Trees and Two-Level AND/OR Circuits.* IEEE Trans. on Computers Vol. 39 (2) (1990), pp.282-287.

[Bru 90]   J. Bruck. *Harmonic Analysis of Polynomial Threshold Functions.* SIAM Journal on Discrete Mathematics Vol. 3 (1990), pp.168-177.

[BS 92]   J. Bruck, R. Smolensky. *Polynomial Threshold Functions, AC^0 Functions, and Spectral Norms.* SIAM Journal on Computing Vol. 21(1) (1992), pp.33-42.

[CDR 86]   S. Cook, C. Dwork, R. Reischuk. *Upper and Lower Time Bounds for Parallel Random Access Machines without Simultaneous writes.* SIAM Journal on Computing Vol. 15(1) (1986), pp.87-97.

[FKPS 85]   R. Fagin, M.M. Klawe, N.J. Pippenger, L. Stockmeyer. *Bounded-Depth, Polynomial Size Circuits for Symmetric Functions.* Theoretical Computer Science Vol. 36 (1985), pp.239-250.

[HMM 82]   S.L. Hurst, D.M. Miller, J.C. Muzio. *Spectral Method of Boolean Function Complexity.* Electronics Letters Vol. 18 (33) (1982), pp.572-574.

[KKL 88] J. Kahn, G. Kalai, N. Linial. *The Influence of Variables on Boolean Functions.* Proc. 29th FOCS (1988), pp.68-80.

[Kar 76] M.G. Karpovsky. *Finite Orthogonal Series in the Design of Digital Devices.* John Wiley and Son, New York (1976).

[KM 91] E. Kushilevitz, Y. Mansour. *Learning Decision Trees using the Fourier Spectrum* Proc. 23rd STOC (1991), pp. 455-464.

[Leh 71] R. J. Lechner. *Harmonic Analysis of Switching Functions.* In *Recent Development in Switching Theory,* Academic Press (1971), pp.122-229.

[LMN 89] N. Linial, Y. Mansour, N. Nisan. *Constant Depth Circuits, Fourier Transform, and Learnability.* Proc. 30th FOCS (1989), pp.574-579.

[Loo 53] L.M. Loomis. *An Introduction to Abstract Harmonic Analysis.* Van Nostrand, Princeton, New Jersey (1953).

[Nis 89] N. Nisan. *CREW PRAMs and Decision Trees.* Proc. 21st STOC (1989), pp. 327-335.

[Raz 92] A.A. Razborov. *On Submodular Complexity Measures.* In "Boolean Function Complexity", Edited by M.S. Paterson, Cambridge University Press (1992), pp.129-139.

[Sim 83] H.U. Simon *A Tight $\Omega(\log \log n)$ Bound on the Time for Parallel RAM's to compute non degenerate Boolean functions.* FCT'83, Lecture notes in Comp. Sci. 158, 1983.

[Weg 87] I. Wegener. *The Complexity of Boolean Functions.* Wiley-Teubner Series in Computer Science. John Wiley and Son (1987).

Graph Theory and Interactive Protocols for Reachability Problems on Finite Cellular Automata

A. Clementi
Department of Computer Science
University "La Sapienza" of Rome - Italy

R. Impagliazzo
Department of Computer Science & Eng.
University of California, San Diego - California - U.S.A.

Abstract

Studying the computational complexity of the Configuration Reachabilty Problem (CREP) is a good way to investigate properties of a given discrete deterministic dynamical system like a Toroidal Cellular Automaton (TCA). We study CREP for two natural *weakly predictable* classes of TCA: the *boolean-disjunctive* class and the *additive* one. For the first class, we reduce CREP to a path problem on a strongly connected digraph and we show a polynomial-time algorithm for this problem. Some consequences of this result on arbitrary TCA are also analysed. For the second class, we show that CREP is not easier than computing the vectorial version of the *Discrete Log Problem* (DLP). However, we also show CREP is unlikely to be NP-complete. To do this, we prove that CREP is in $Co - AM[2]$, where $AM[2]$ is the class of problems with a constant round interactive protocol [1, 13]. CREP is unlikely to be NP-complete (unless the polynomial time hierarchy collapses), then follows by the results of [3]. All such results hold even when multidimensional and/or non homogeneous TCA arise. As a global consequence, we argue that the structure of the weakly predictable class resembles the NP one.

1 Introduction and motivations

Cellular automata were introduced by J. Von Neumann as a universal computational model satisfying physical laws. Now, the reason of the growing interest in the theory of cellular automata can be found in the fact that dynamical systems, with "chaotic" (that is unpredictable) behaviour, such as parallel and distributed computer systems, fluidodynamic systems, neural networks and biological systems, have "easy" elementary units, local "instantaneous" interactions and simple local rules of evolution. Thus, the "nasty" behaviour comes from the interaction of a large number of elements rather than the "hardness" of some particular local law. From this point of view, cellular automata are an universal mathematical model capturing such a phenomenon. A *D-Dimensional Toroidal (homogeneous) Cellular Automaton* (also denoted as TCA) is a triple $A = (Q, N, f)$ over the support $\mathbf{T} = (\mathcal{Z}^D_{modn}, +)$, that is, a toroidal grid of n^D identical finite automata (cells) with state set Q and locally connected each other. The state of each cell depends on the state of the neighbouring cells (determined by the set of **T**-vectors N) and varies according to the the local rule $f : Q^{|N|} \rightarrow Q$. Concerning the global evolution, the cells change their states in a discrete synchronous way and f uniquely determines a global function F acting on the space $\Sigma = Q^{n^D}$ of all possible configurations. Thus, the behaviour of the system can be described by the following equation:

$$X^{t+1} = F(X^t) \qquad X^t \in \Sigma \qquad t \geq 0 \tag{1}$$

With this definition all cells have the same neighbourhood and have the same local function. However, it is also interesting to consider *non-homogeneous* cellular automata (in which such restrictions are relaxed) since any homogeneous multidimensional TCA ($D > 1$) can always be seen as an one dimensional non-homogeneous TCA (for the reduction and further discussions see section 2). Moreover, some applications of *non-homogeneous* TCA deal with the possibility to define one-way functions (hard to invert) in a more powerful way than in the homogeneous case ([10, 26, 24]).

Interesting recent studies have been made on the theory of cellular automata [25, 21, 7], nevertheless, important questions still remain open. Wolfram [23] defined a list of crucial problems in the study of cellular automata and we shall discuss two of them. The first problem is the definition of an overall classification of cellular automata based on a "significant" complexity measure. Partial answer to this question have been recently given ([22, 7, 8, 11]): on the one hand, concept arising from the theory of dynamical systems have been used to characterize some properties of cellular automata such as entropy and speed of information propagation ([22, 11]); on the other hand, the temporal evolution of a given cellular automaton has been seen as a parallel computation, so that, the nature of the evolution has been figured out using methods from computation and formal language theory.

The second problem defined by Wolfram is: how common is *unpredictability* in cellular automata? That is, in order to determine the behaviour of a given cellular automaton, are we forced to simulate (step by step) its evolution or we can find a good "shortcut" to predict its future states?

A uniform analysis of both questions can start from the strong connections existing between the *computational power* (or the behaviour complexity) of a fixed cellular automaton class **C** and the complexity of the following problem:

- *CREP (Configuration Reachability Problem)*

  Instance: a cellular automaton $A \in \mathbf{C}$ with global function F and two configurations X, Y.

  Question: does there exist t such that $F^t(X) = Y$?

These connections, which are intuitively clear, have been deeply discussed for example in [5, 23].

Let us now focus our attention on CREP. Of course, this problem is decidable for any finite cellular automaton (TCA) since the space Σ is a finite set. However, its computational complexity varies greatly depending on dimension, local function and neighbourhood structure of the TCA considered in input. This suggests that the computational complexity of CREP can play the role of a significative measure of complexity for TCA. It is well-known that CREP is *PSPACE*-complete for arbitrary local maps even when the dimension is 1 (see for example [21]). Thus, it seems that there is no "shortcut" to solve CREP other than generating successively all configurations in the orbit of the starting configuration. Moreover, let us consider a slight variant of CREP, that is, given $t \geq 0$ and a configuration X, compute $F^t(X)$. From the *PSPACE*-completeness of CREP it follows that the latter problem is not solvable in polynomial time, that is, in time $O[(log t)n^k]$ (where n is the period (size) of the toroidal support and k is some constant positive integer). Let us then call a TCA *weakly predictable* if we can compute $F^t(X)$ in polynomial time. Moreover, we say that a TCA is *predictable* if CREP is decidable in polynomial time. Weakly predictable and predictable notions formalize the concept of *computationally reducible* dynamical systems (see also [5, 23]); in particular, the class of predictable TCA can be seen as the finite version of the class of infinite cellular automata for which CREP is decidable (see the Culik and Yu classification [9]).

In this paper, we analyse the complexity of CREP for two natural classes of weakly predictable TCA.

The first class, also called *boolean-disjunctive*, is a non-trivial example of predictable class. For these automata, the state set Q is the binary alphabet and the local function f is a "disjunction"[1] of an arbitrary number of neighbours. In this case, we reduce CREP to a "path" problem in a strongly connected digraph and define a polynomial-time algorithm working also for multidimensional and/or non-homogeneous TCA. An interesting consequence of this result, for general TCA, is the following. Let us consider a TCA as a parallel synchronous network of "elementary processors" in which we can specify the input and output informations by making use of two fixed subset of cells (input and output zone). The *PSPACE*-completeness of CREP states that, in general, it is hard to predict whether a *specific* input yelds after a certain time t a *specific* output or not. Our results, however, allow to efficiently decide whether the input and the output zones can "communicate" *arbitrary* informations within a certain time t; indeed, the communication structure of any TCA can be represented by a boolean-disjunctive TCA. Thus, if the answer to the latter problem is negative, we can certainly state that CREP has always negative answer.

The second class we consider is the *additive* one. We know that, for weakly predictable TCA, CREP is in NP, (indeed, it is NP-complete in dimension greater than one (see [21]) and still unknown in one dimension). Since the existence of a class of weakly predictable TCA which are not predictable is equivalent to the assertion $P \neq NP$, it should be very difficult to find a description of such a class.

---

[1]that is f is the OR function

In this research direction, one of the most considered class (see for example [18, 12]), containing weakly predictable TCA, is the *additive* class: the state set is a finite field and the local map is a linear combination of the states of neighbour cells. In section 4, we study CREP for the additive class. First of all, we show that such a problem is not easier than computing the vectorial version of the *Discrete Log Problem* (DLP) in a finite field with exponential cardinality with respect to the size of the TCA considered in input. Since the vectorial version of DLP is widely believed to be computationally difficult (see [20]), this suggests that no polynomial-time algorithm exists for our problem. However, we also show that CREP cannot be NP-complete even in the multidimensional and/or non-homogeneous cases unless the polynomial-time hierarchy collapses. In order to do this, we prove that CREP is in $CoAM[k]$, where $AM[k]$ is the class of problems having an interactive protocol with a constant number of interactions ([1, 13]).

2 Preliminaries

Let us give the definition of the boolean-disjunctive class and the additive one showing the common aspects. We first consider homogeneous 1-dimensional TCA, then we shall define the general case.

A 1-dimensional TCA $A = (Q, N, f)$ consists of a line of n sites ($i = 0, \ldots, n-1$), arranged around a circle (so as to give periodic boundary conditions) and with each site carrying a value of the set Q. The dynamical evolution of A is determined by f in the following way[2]:

$$X_i^t = f(X_{i+j_0}^{t-1}, \ldots, X_{i+j_h}^{t-1})$$

where the neighbourhood is $N = (j_0, \ldots, j_h)$ ($j_k \in Z_n$) and X_i^t is the state of the cell i at time t.

Definition 2.1 Boolean-disjunctive TCA *A boolean-disjunctive TCA $A = (Q, N, f)$ is defined as : $Q = \{0, 1\}$; $N = (j_0, \ldots, j_h)$ (($j_k \in Z_n$)); and f :*

$$f(X_{i+j_0}, \ldots, X_{i+j_h}) = \bigvee_{k=0}^{h}(C_k \wedge X_{i+j_k}) \qquad C_k \in Q \quad k = 0, \ldots, h$$

Definition 2.2 Additive TCA *An additive TCA $A = (Q, N, f)$ is defined as: $Q = GF(p) = \langle Z_p, +, * \rangle$ for some prime integer p; $N = (j_0, \ldots, j_h)$ ($j_k \in Z_n$); and f:*

$$f(X_{i+j_0}, \ldots, X_{i+j_h}) = \sum_{k=0}^{h}(C_k * X_{i+j_h}) \qquad C_k \in Z_p \quad k = 0, \ldots, h$$

We observe that, in both classes, the complete description of a TCA of size n can be simply given by defining the following $n \times n$ matrix C:

- *Boolean-disjunctive TCA:* C is a boolean matrix and $C(i, j) = 1$ if there exists $j_k \in N$ such that $j = i + j_k$ and $C_k = 1$; $C(i, j) = 0$ otherwise.

---

[2]in which follows, all index operations have to be performed *mod n* and all non-specified indexes will be in the range $\{0, \ldots, n-1\}$.

- *Additive TCA:* $C \in (GF(p))^{n \times n}$ and $C(i,j) = C_k$ if there exists $j_k \in N$ such that $j = i + j_k$; $C(i,j) = 0$ otherwise.

Note that, the class of homogeneous TCA will correspond to the particular subclass of *circulant* matrices, that is, the matrices in which $C(i,j) = C(i+1,j+1)$ for any $i, j = 0, \ldots, n - 1$.

Concerning the multidimensional case $(D > 1)$, we can easily transform any D-dimensional TCA of size n into a 1-dimensional TCA of size n^D. Indeed, without loss of generality[3], consider the boolean-disjunctive case in which $D = 2$; in this case, we can define the matrix $C \in \{0, 1\}^{n^2 \times n^2}$ and, thus, the corresponding 1-dimensional boolean-disjunctive TCA A, where, for $i, j = 1 \ldots n^2$, $C(i,j) = 1$ if the cell $Ind^{-1}(j)$ is a neighbour of $Ind^{-1}(i)$. Here, $Ind : \{0, \ldots, n - 1\}^2 \rightarrow \{1, \ldots, n^2\}$ is any fixed one to one pairing function.

Actually, with this reduction, the matrix C does not keep the topological properties of the associated multidimensional TCA. Even if we consider a simple 2-dimensional neighbourhood like the *Moore* one[4] the corresponding matrix C seems to have no interesting properties and A becomes non-homogeneous. This is one of the crucial reasons for which, in general, several problems, for arbitrary homogeneous cellular automata, drastically increase their complexity when the multidimensional case arises (see [16]).

Let us conclude this section by giving the appropriate definition of CREP for boolean-disjunctive and additive TCA. Given a TCA A uniquely defined by a matrix C and a starting configuration X, the next configuration Y can be computed in the following way:

$$Y = F(X) = CX$$

where each component of Y (state of cell i, $i = 0, \ldots, n - 1$) is:

1. $Y_i = \bigvee_k (C(i,k) \wedge X_k)$ (boolean-disjunctive TCA)

2. $Y_i = \sum_k (C(i,k) * X_k)$ (additive TCA)

In both cases, the operations have the associative and distributive properties, then, for any $t > 0$:

$$X^t = F^t(X) = C^t X$$

that is, we can figure out the t-th configuration in the orbit starting from X by simply computing the t-th power of C. Thus, the following facts are easy consequences of the above arguments:

Lemma 2.1 *Given a boolean-disjunctive (or additive) TCA of size n, a configuration X and $t > 0$, it is possible to compute X^t in polynomial time; hence, boolean-disjunctive and additive TCA are weakly predictable. Moreover, CREP (which is in NP for weakly predictable TCA) can be defined in the following way:*

---

[3] the same procedure can be applied for the other cases ($D > 2$ and additive TCA)
[4] The Moore neighbourhood, for 2-dimensional supports, consists of the eight surrounding cells of the generic cell

- Instance: *a matrix* $C \in \{0,1\}^{n \times n}$ *($C \in (GF(p))^{n \times n}$ - for additive TCA); two configurations X,Y;*

- Question: *Does non negative integer t such that $C^t X = Y$ exist?*[5]

3 The boolean-disjunctive class

We are going to prove that, in this case, CREP can be "efficently" solve using notions from graph theory.

As we have seen in the previous section, any boolean-disjunctive TCA can be defined by a boolean matrix C. C can be seen as the adjacency matrix of a digraph $G(V, E)$ of n vertex and CREP is equivalent to the following problem:

- $CREP^G$

- *Instance:* A digraph $G(V, E)$ and $S, P \subseteq V$.

- *Question:* Does non negative integer t exist such that for every $s \in S$ there exists $p \in P$ such that there exists a *walk* $w_{s,p}$ from s to p with length t and, furthermore, P is maximal (that is there is no $p' \in V - P$ for which there exists a walk of length t from some $s \in S$ to p') ?

For *walk* we shall always intend a generic path in which one edge can be eventually considered more than one time. Informally, given a graph $G(V, E)$ and two sets of node S and P we ask whether it is possible to connect every S-node to a P-node by walks of the same length and this property does not hold for any node not in P, for that length.

To show the equivalence between CREP and $CREP^G$ we simply associate with any boolean-disjunctive TCA to the above defined graph G and for any pair $(X, Y) \in \{0, 1\}^n \times \{0, 1\}^n$ we consider, respectively, the vertex subsets P and S in which $i \in S$ ($i \in P$) if and only if $X_i = 1$ ($Y_i = 1$). It is easy to prove that such a reduction is bijective.

Note that, in general, $CREP^G$ cannot be reduced to the classical transitive closure of a digraph since the required paths are not necessarily simple, so their length can be greater than n.

In this section we study the complexity of CREP when the TCA considered are not in general homogeneous, that is, we relax the restriction in which every cell has the same neighbourhood. More precisely, we analyse the case in which the corresponding graph G (associated with the TCA) is strongly connected. This new restriction, although in graph theory is a "significant" one, when we deal with graphs associated with TCA it is motivated by the practical necessity that every pair of "processors" in an efficient network must be able to communicate each other (directly or not).

Let us denote as *cycle* any cyclic walk not passing on the same edge more than one time; we define $d(G)$ as follows:

---

[5]note that this lemma holds also for non-homogeneous and/or multidimensional TCA

$$d(G) = g.c.d.\{|C| \; : \; C \text{ is a cycle of } G\} \tag{2}$$

The function $d(G)$ can be computed in polynomial time:

Lemma 3.1 *Let FP be the class of polynomial-time computable functions, then, for any strongly connected graph G, $d(G) \in FP$ (more precisely $d(G) \in O(n^3)$).*

Proof. Let us denote as C the adjacency matrix of G and consider the first $n + 1$ powers of C where the operation $C^t = C^{t-1}C$ is the classical one:

$$C^t(i, j) = \bigvee_{k=1}^{n} (C(i, k)^{t-1} \wedge C(k, j)) \qquad t = 2, \dots, n+1$$

It is easy to see that $C^t(i, j) = 1$ if and only if there exists a (at least one) walk from i to j of length exactly t. Now, we consider the set of "lengths" T defined as follows:

$$T = \{t : t \leq n + 1 \text{ and there exists } i \text{ such that } C^t(i, i) = 1\}$$

then, we consider $d = g.c.d.(T)$; note that all cycle lengths $|C|$ can be written as $|C| = |K_1| + .. + |K_c|$ where K_i is a *simple* cycle of length not greater than $n + 1$ (not visiting the same vertex more than one time) and, moreover, d must devide K_i for every i. Hence, from definition (2), it is not hard to show that d is equal to $d(G)$. □

Let us consider the case $d(G) > 1$ (case $d(G) = 1$ will be implicitly analysed in the proof of lemma 3.3)

Lemma 3.2 *For any strongly connected graph G and for any pair (s, p) of nodes in G, there exists an integer $k_{s,p}$ $(k_{s,p} < d(G))$ (depending only on (s, p)), such that the length $l_{s,p}$ of every possible simple path from s to p satisfies:*

$$l_{s,p} = k_{s,p} \; mod \; d(G)$$

In other words, we can choose any simple path from s to p, when we deal with properties modulus $d(G)$.

Proof. Let us assume that we have two simple paths $p_{s,p}$ and $p'_{s,p}$ with (different) lengths denoted, respectively, as $l_{s,p}$ and $l'_{s,p}$ and consider a path (it always exists) from p to s denoted as p^{-1}, then by the definition of $d(G)$ we have:

$$l_{s,p} + |p^{-1}| = 0 \; mod \; d(G)$$

and

$$l'_{s,p} + |p^{-1}| = 0 \; mod \; d(G)$$

Since, for any positive integer d, the set of integers $mod \; d$ is an additive group, the result is proved.

□

Lemma 3.3 *With the same definitions of lemma 3.2, every walk $w_{s,p}$ from s to p has a length satisfying:*

$$|w_{s,p}| = k_{s,p} \bmod d(G)$$

Moreover, given two arbitrary pairs of nodes (s,p) and (s',p') if $k_{s,p} = k_{s',p'} \bmod d(G)$, then there exists t such that there exist two walks w and w', respectively from s to p and from s' to p', whose lengths are both equal to t.

Proof. Any walk from s to p has a length satisfying the following formula:

$$|w_{s,p}| = l_{s,p} + h_1 C_1 + \ldots + h_w C_w \ (h_i > 0 \ i : 1, \ldots, w)$$

where $l_{s,p}$ is the length of a simple path from s to p and C_i ($i : 1, \ldots, w$) are the lengths of some simple cycles of G. Taking the $d(G)$-modulus of both sides of the above formula and using lemma 3.2, the first part of lemma holds.

For the second part we are going to use the following well-known result in number theory (see for example [15]).

Theorem 3.1 *For any finite set of integers $\{C_1, \ldots, C_k\}$, such that*

$$g.c.d.(\{C_1, \ldots, C_k\}) = 1$$

there exists a positive integer B_0 such that for any integer $B \geq B_0$ there exist $h_1, \ldots, h_k$ ($h_i \geq 0 \ i = 1, \ldots k$) satisfying the following expression:

$$B = h_1 C_1 + \ldots + h_k C_k$$

We now prove that, when $d(G) = 1$, for any choose of (s,p) and (s',p') we can always find two walks w and w', respectively from s to p and from s' to p', having the same length. In fact, there are at least two walks w and w' satisfying the following expressions:

$$|w| = l_{s,p} + h_1 C_1 + \ldots + h_k C_k$$
$$|w'| = l_{s',p'} + h_1' C_1 + \ldots + h_k' C_k$$

where all values have been defined above and, moreover,

$$g.c.d.\{C_1, \ldots, C_k\} = 1$$

By theorem 3.1, there exists a positive integer B_0 such that, choosing w such that $|w| \geq B_0$ (we can always do this since G is strongly connected), we can set $h_1', \ldots, h_k'$ in order to have $|w| = |w'|$.

Let us now consider the case $d(G) > 1$. In such a case, using previous notations, we claim that if we have two pairs (s,p) and (s',p') such that $k_{s,p} = k_{s',p'} \bmod d(G)$ then, for infinitely many t, there exist two walks w and w' respectively from s to p and from s' to p' having lengths both equal to t. In fact, for any choose of w and w', the following equations hold:

$$|w| = k_{s,p} + \alpha d(G) + h_1(d(G)C_1) + \ldots + h_k(d(G)C_k)$$

$$|w'| = k_{s,p} + \alpha' d(G) + h_1'(d(G)C_1) + \ldots + h_k'(d(G)C_k)$$

where $g.c.d.(\{C_1, \ldots, C_k\}) = 1$.

Considering both the equations $mod\ d(G)$ we can now repeat the same procedure of the case $d(G) = 1$. □

As a global consequence of the last results, it is possible to prove the following proposition:

Theorem 3.2 $CREP^G$ *is decidable in time $O(n^3)$ for strongly connected digraph. Hence, boolean-disjunctive TCA are predictable even in the non-homogeneous and/or multidimensional cases.*

Proof. Consider the set $S \times P$, where S and P are the sets of the $CREP^G$ instance; for each pair $(s, p) \in S \times P$ we consider a simple path from s to p and compute the integer $k_{s,p}$ ($k_{s,p} < d(G)$). Now, for any $d' \in \{0, \ldots, (d(G)-1)\}$, we check whether for any $s \in S$ there exists (at least one) $p \in P$ such that $k_{s,p} = d'$ and whether P is a maximal set for such a property (see the definition of $CREP^G$). If we find at least one d' satisfying the last property then we accept the instance otherwise we reject. From lemma 3.3, it is easy to verify that the procedure correctly decides the existence of walks of the same length from each $s \in S$ to some $p \in P$.

The polynomial time ($O(n^3)$) of the global procedure is a consequence of lemma 3.1, the relation $d(G) \leq n$ and the fact that $|S \times P| \in O(n^2)$. Moreover, the above procedure does not depend on whether C (that is, the adjacency matrix of G) is circulant or not, thus it can be applied even for non-homogeneous and/or multidimensional TCA. Finally, by the equivalence between CREP and $CREP^G$ (for boolean-disjunctive TCA), the theorem is proved.

□

4 The additive class

In this section we consider the class of *additive* cellular automata. Here, the alphabet is a finite field $GF(p) = \langle Z_p, +, * \rangle$ for some prime integer p and the modality of evolution is determined by[6]:

$$X_i^t = \sum_k (X_k^{t-1} * C_{i,k}) \tag{3}$$

where $C_{i,k}$ is in Z_p.

If we write the above equation in vectorial form we obtain:

$$X^{t+1} = CX^t \qquad X^t = C^t X^0 \tag{4}$$

---

[6] as in the previous section, we shall use the convention that all non specified indexes will be in the range $\{0, \ldots, (n-1)\}$ and, moreover, all index operations have to be performed $mod\ n$)

where $C \in (GF(p))^{n \times n}$ and $X^t \in (GF(p))^n)$ for any $t \geq 0$.

It is not hard to verify that the CREP is in NP, since we can "guess" t and compute $C^t X$ in polynomial time ($O(n^3 \log t)$).

Any matrix $C \in GF(p)^{n \times n})$ can be transformed into an equivalent one in Jordan form (denoted as C^J). More precisely, consider the n-th primitive roots $\xi_0, \ldots, \xi_{n-1}$ of the unity in $GF(p)$, then the elements of C^J vary over all the smallest field containing such roots. This field is called the *splitting field* of $GF(p)$ by $x^n - 1$ and it is isomorphic to the field $GF(p^m)$ where m is the *multiplicative order* of p modulus n (i.e. the least integer m such that n devides p^m). From a complexity point of view, we observe that m can be of the same order of n, in fact, this holds when n is prime and p is a generator of $GF(n)$ (that is, each element of the field can be expressed as a power of p). We can summarize these notions in the following way (for an exhaustive analysis of this part see [19]):

Theorem 4.1 *Any $n \times n$ matrix C with elements in $GF(p)$ is equivalent (isomorphic) to a Jordan matrix C^J with elements in $GF(p^m)$ and the diagonal elements λ_k are the eigenvalues of C.*

Note that we can compute the matrix J (i.e. the Jordan base) for C such that $C^J = J C J^{-1}$ - thus, all the eigenvalues of C - in polynomial time (see also [12]).

Let us now consider homogeneous TCA; in this case C is circulant (see section 2), that is $C_{i,k} = C_{i+1,k+1}$, so C can be described by a n-vector $C = (c_0, \ldots, c_{n-1})$ where $c_i \in GF(p)$. When C is circulant it is possible to prove (see [12]) that the eigenvalues of C are:

$$\lambda_k = \sum_i c_i \xi_k^i \tag{5}$$

Moreover, when $g.c.d.(p, n) = 1$ the Jordan form C^J is always a diagonal matrix with the elements on the principal diagonal equal to the eigenvalues of C. Under this restriction, we are able to show that CREP is equivalent to the vectorial version of the Discrete Log Problem (DLP) in the field $GF(p^m)$. Indeed, in such a case CREP is equivalent to decide whether there exists t such that the following equation system holds:

$$\lambda_k^t = Y_k X_k^{-1} \qquad k = 0, \ldots, n - 1 \tag{6}$$

The crucial point is that the triple (λ_k, X_k, Y_k) can take all values in $GF(p^m)$ depending on the values of the CREP instance (i.e. the vectors $C = (c_0, \ldots, c_{n-1})$, X and Y). To see this we simply observe that since $\{\xi_k : k = 0, \ldots, n - 1\}$ is a set of generators for $GF(p^m)$ then our claim is a direct consequence of the expression 5 and of the fact that the Jordan forms X^J and Y^J of the configurations X and Y, satisfy, respectively, the equations $X_k^J = \sum_i X_i \xi_k^i$ and $Y_k^J = \sum_i Y_i \xi_k^i$.

However, the algebraic characteristics of the additive class (homogeneous and not) can be strongly used to study the properties of the state transition diagram of such systems. Let us consider an additive TCA A uniquely determined by its matrix C ($C \in GF(p^{n \times n})$) and consider the Jordan form $C^J = J C J^{-1}$. The *state transition diagram* of A is the digraph $\mathcal{D} = (V, E)$ where V is the sets of all possible configurations X of A (i.e. $V = \{0, \ldots, p-1\}^n$) and $(X, Y) \in E$ if $Y = CX$. Note that, in general, $\mathcal{D}$ is not connected and since the rule is deterministic, each vertex has a unique successor; since the number of possible configurations ($|V|$) is finite, the system starting from any initial state must always reach a cycle. Hence, $\mathcal{D}$ consists of cycles (which may be points) with trees rooted on the cycle vertices. Using notions from linear algebra it is possible to prove the following results (see [18] or [12]):

Lemma 4.1 *For any additive cellular automaton A, we have:*

- *A is reversible (that is, each vertex of $\mathcal{D}$ is on a cycle) if and only if C has not the eigenvalue 0, that is, if and only if $det(C) \neq 0$;*

- *Suppose A is not reversible; X will be on a cycle if and only if its corresponding Jordan form $X^J = JXJ^{-1}$ has no component on the eigenspace S_0 associated with the eigenvalue $\lambda = 0$ of C.*

- *if X is not in a cycle of $\mathcal{D}$ the length l of the path (unique) starting from X and reaching the first vertex X^c on a cycle (i.e. the length of the transient part of the X orbit) is bounded by the dimension k_0 of S_0, hence, $l \leq n$.*

- *Every cycle of $\mathcal{D}$ has a length c which is a factor of $p^m - 1$.*

4.1 An Interactive Protocol for CREP

The principal goal of the previous analysis is to give significative motivations for our next non "standard" algorithm for CREP in the additive class. Indeed, we have seen that the existence of a deterministic polynomial-time algorithm for CREP in the additive class would imply a rather surprising result in number theory. In the following, we shall give some definitions and results about the theory of interactive probabilistic protocols (see [2, 13]).

The class NP can be defined as the set of languages L for which there exists a proof checkable in deterministic polynomial time for any string $x \in L$ and, on the other hand, no proof (and so no prover) can be found for $x \notin L$. For example, considering the well known NP-complete language SAT and a boolean formula f in conjunctive normal form (i.e. an instance for SAT); an "efficient" proof for $f \in SAT$ consists of any truth assignment satisfying f. If f is not satisfiable, then no assignment satisfying f (that is, no proof) can be given. The theory of efficient computing shows, however, that random can increase power, provided that we allow a small probabilty of error. Applying this idea to provability Babai ([1]) and Micali et al ([13]) proposed two equivalent randomized generalizations of NP (for the proof of the equivalence see [14]). The focus idea is in the notion of a "proof" as an interaction between two players: a *Prover* and a *Verifier*. Players interact by sending messages to each other and after a polynomial number of interactions (in the size of input), *Verifier* performs a polynomial-time randomized computation to determine whether he is convinced that $x \in L$. Thus, *Verifier* is a probabilistic Turing machine working in polynomial time with respect to the input length. On the other hand, *Prover* is a computationally unlimited (both in time and space) deterministic Turing machine. Moreover, *Verifier* and *Prover* take turns in being active and the former starts the computation. When a machine is active, it can perform internal computations, read and write on the proper tapes and send a message to the other machine on an appropriate communication tape. The length of messages exchanged between the machines is bounded by a suitable polynomial in the input length. Finally, *Verifier* can, during its turn, terminate the interactive computation by entering in a final (accepting or rejecting) state. The acceptance cryterion is equivalent to that of BPP-machines[7]:

Definition 4.1 *a language L has an interactive protocol if, for any $x \in L$, there exists a* Prover *such that* Verifier *accepts x with probability greater than 2/3. On the contrary, if $x \notin L$,* Verifier *accepts x with probability smaller than 1/3 for any* Prover.

Furthermore, we define $AM[k(n)]$ the class of all languages for which there exists an interactive protocol using at most $k(|x|)$ interactions (rounds) between players for any input x.

Now, we give some results which are crucial for our pourpose:

Theorem 4.2 [2] $AM[k] \equiv AM[2]$ *for any fixed constant k, and $AM[2] \in PH$ where PH is the polynomial-time hierarchy (more precisely, $AM[2] \in \Pi_2^p$)[8].*

Theorem 4.3 [3] *If $CoNP$ is contained in $AM[k]$ for some constant k, then PH is contained in $AM[k]$, that is PH collapses.*

In the following, we construct an interactive protocol with a constant number of rounds for the complement of CREP; thus, we shall obtain the same result shown for the well known graph isomorphism problem (GIP) (see for example [17]).

Let us consider the case in which $det(C) \neq 0$ (that is C is invertible). The general case will be analysed in the second part of this section.

Given any $C \in GF(p)^{n \times n}$ such that $det(C) \neq 0$, it is not hard to show that the set of powers of C:

$$G = \{D = C^t : \text{ for some } t \geq 0\}$$

satisfies the following algebraic properties:

Lemma 4.2 *Let us denote the identity matrix in $GF(p)^{n \times n}$ as I and the least positive integer t such that $C^t = I$ as $ord(G)$, then:*

a) $C^{-1} = C^{(ord(G)-1)}$;

b) G is a multiplicative group of $GF(p)^{n \times n}$ where $(C^t)^{-1} = (C^{-1})^t$;

---

[7] A more detailed definition of this computational model can be found in [4]

[8] For the definition of PH see for example [4]

Let us consider two arbitrary configurations X and Y; since G is a group, if $DX = Y$ for some $D \in G$ then there exists $D' \in G$ such that $D'Y = X$; thus, the relation in $GF(p)^n$ defined as:

$$X \equiv_G Y \text{ if there exists } D \in G \text{ such that } DX = Y$$

is an equivalence one.

From this definition, it is easy to verify that CREP for reversible additive TCA is equivalent to, giving G and X, Y in $GF(p)^n$ as input, decide whether $X \equiv_G Y$.

Now, we give the interactive protocol for the complement of CREP:

- *Protocol A1*

- *Input:* $C \in GF(p)^{n \times n}$ and $X_1, X_2 \in GF(p)^n$

- *Verifier* chooses both uniformly at random $i \in \{1, 2\}$ and $t \in \{0, \ldots, p^m\}$;

- *Verifier* computes the matrix $C^t \in G$ and $C^t X_i = Z$;

- *Verifier* sends Z to the *Prover* and asks to him whether $X_1 \equiv_G Z$ or $X_2 \equiv_G Z$;

- *Prover* replies to *Verifier* by sending the index i';

- *Verifier* checks whether i' is correct (i.e. if $i' = i$).

- *Verifier* will accept the *Prover* answer if and only if i' is correct.

Theorem 4.4 *The protocol* A1 *works correctly (that is, it satisfies the conditions given in definition 4.1).*

Proof. The protocol is *efficient*; all computations done by *Verifier* can be realized in time $O(n^3 \log t)$ by a probabilistic Turing machine and the number of interactions is constant for any size n of the input. Moreover, the message length is bounded by n.

The protocol is *complete*; if X_1 is not equivalent to X_2 and both *Verifier* and *Prover* follow the protocol then *Verifier* will accept with probability higher than 2/3 (in this case with probability 1). So there exists at least one strategy of *Prover* which convinces *Verifier* to accept with "high" probability.

The protocol is *sound*; Let us define the following sets:

$$P_Z^i = \{D : D \in G \quad and \quad DX_i = Z\} \quad i = 1, 2 \tag{7}$$

Since G is a group it is not hard to prove that if $X_1 \equiv_G X_2$ then $|P_Z^1| = |P_Z^2|$. For this fact and for the fact that *Prover* cannot see index i, we can claim that for any strategy of *Prover* the probabilty that *Verifier* accepts is not greater than 1/2 when $X_1 \equiv_G X_2$. Hence, with two independent runs of the protocol on the same input (that is, with a total number of interactions equal to 4), we satisfy the second condition of definition 4.1. $\square$

We conclude this section showing how to extend the protocol *A1* to the case $det(C) = 0$ (that is C is not invertible). From lemma 4.1, if $det(C) = 0$ then

$\lambda_0 = 0$ is an eigenvalue of C and the dimension k_0 of the corresponding eigenspace is greater than 0. Without loss of generality, we can assume that the Jordan form C^J has the S_0-block as the first Jordan block (the first k_0 rows). Moreover, the semigroups G and $G^J = \{D = (C^J)^t : \text{ for some } t \geq 0\}$ are isomorphics since if C and D are matrices in $GF(p)^{n \times n}$ then $(C * D)^J = C^J * D^J$, thus for any $t > 0$:

$$(C^J)^t = (C^t)^J$$

Given two configurations X, Y (with their Jordan forms respectively denoted as X^J and Y^J) four cases may arise:

i) X^J and Y^J have both non-zero components in subspace S_0; in such a case from lemma 4.1, both configurations are in the transient part[9] of some orbits of the system, thus, we can check, computing

$$CX, C^2 X, \dots, C^{k_0} X$$

whether X can reach Y and the computation is polynomial in n ($O(k_0 n^2)$) (note that $k_0 \leq n$).

ii) X^J has no component in the subspace S_0 and, on the contrary, Y^J has non-zero components in S_0; Since, from lemma 4.1, X is in a cycle (its orbit never leaves such a cycle) and Y is not in a cycle we claim that X cannot reach Y.

iii) X^J and Y^J have both no component in subspace S_0; let us consider the $(n - k_0) \times (n - k_0)$ submatrix E of C^J obtained erasing the first k_0 rows and columns of C^J; E is an invertible matrix and, in this case, X^J and Y^J can be considered elements of $GF(p)^{(n-k_0)}$. Thus, we have reduced our initial instance (C, X, Y) to an equivalent one (E, X^J, Y^J) in which $det(E) \neq 0$ and so we can apply the interactive protocol $A1$.

iv) X^J has non-zero components in the subspace S_0 and, on the contrary, Y^J has no component in S_0; In this case Y is on a cycle, so, in time $O(k_0 n^2)$, we can compute, by lemma 4.1, the configuration X^c (and its Jordan form) which is the first one in the cycle reachable from X. Finally, we can repeat the same procedure of case *iii)*.

As final results of this section we have:

Theorem 4.5 *CREP for additive cellular automata is in $Co-AM[2]$ even for non-homogeneous and/or multidimensional cellular automata. Thus, CREP cannot be NP-complete unless the polynomial-time hierarchy PH collapses.*

Proof. By the above discussion we know how to reduce the general case to the case in which the matrix C associated with the cellular automaton is invertible; at this point, we can apply the interactive protocol $A1$ to the complement of CREP. This proves that CREP is in $Co-AM[4]$, but by theorem 4.2 this is equivalent to prove

---

[9]that is, both of them are not in a cycle

that CREP is in $Co - AM$ [2]. Moreover, we observe that all results do not depend on the fact that C is circulant or not, thus, the global algorithm can be applied even when non-homogeneous and/or multidimensional cellular automata are considered (see also section 2). The last part of the theorem is a direct consequence of the first one and of theorem 4.3. □

5 Conclusions

An overall interpretation for all results presented in this paper confirms that the approach proposed in [5, 23, 21] and based on the analysis of decision problems like CREP, can give interesting consequences in the theory of cellular automata. Let us summarize such aspects.

Our polynomial-time algorithm for Boolean-disjunctive rules, besides determining a predictable class, gives a tool for analysing the communication structure of arbitrary rules when we consider TCA as parallel computing networks.

The lower bound and the interactive protocol deciding CREP for additive rules show that the additive class does not represent the behaviour of weakly predictable TCA. That is, there must be some weakly predictable TCA having a more "complex" evolution than the additive one, unless $P = NP$. Hence, the additive subclass in the weakly predictable class seems to play the same role of NP-$Intermediate$ (NPI) problems - like the Graph Isomorphism problem (GIP) ([17]). Thus, we can observe (see the following figure) that the structure of the weakly predictable class resembles the NP one. This may be interpreted as a good reason to consider the computational complexity of CREP as a significant complexity measure for finite cellular automata. A future interesting work should be in exploring the concept of reducibility on the ground of this complexity measure and the consequent notion of $complexity - equivalent$ cellular automata (see also [6]).

NP   Weakly Predictable TCA

Comparing NP with the Weakly Predictable Class

Acknowledgements

We would like to thank Prof. P.D. Bovet, Prof. P. Crescenzi and Prof. C.H. Papadimitriou who permit us to work together and Prof. P. Mentrasti, Prof. A. Morgana and Doct. P. Pierini for helpful and very interesting discussions.

References

[1] Babai L.: Trading group theory for randomness. Proc. of STOC, 421-(1985).

[2] Babai L., Moran S.: Arthur-Merlin games: a randomized proof system and a hierarchy of complexity classes. J. of Computer and System Sciences, 36, 254-276 (1985).

[3] Boppana R.B., Hastad J. and Zachos S.: Does Co-NP have short interactive proofs? Information Processing Letter, 127-132 (1987).

[4] Bovet D.P., Crescenzi P. (1993), *Introduction to the theory of computational complexity*, Prenctice Hall.

[5] Buss S., Papadimitriou C.H., Tsitsiklis J.N.: On the predictability of coupled automata: an allegory about chaos. Proc. of FOCS, 788-793 (1990).

[6] Clementi A.: *On the Complexity of Cellular Automata*, Ph.D. Thesis, in preparation.

[7] Culik K., Hurd P. and Yu S.: Formal languages and global cellular automata behaviour. Physica D, 45, 396-403 (1990).

[8] Culik K., Hurd P. and Yu S.: Computation theoretical aspects of cellular automata. Physica D, 45, 367-378 (1990).

[9] Culik II K, Yu S.: Undecidability of CA classification scheme. Complex Systems, 2 (2), 177-190 (1988).

[10] Guan P.: Cellular automata public key cryptosystem, Complex Systems, 1, 51-57 (1987).

[11] Gutowitz H.A.: A hierarchycal classification of cellular automata. Physica D 45, 136-156 (1990).

[12] Guan P., He Y.: Exact results for deterministic cellular automata with additive rules. J. of Stat. Physics, 43, 463-478 (1986).

[13] Goldwasser S., Micali M. and Rackoff T.: The knowledge complexity of interactive proof systems. SIAM J. of Computing, 18, 186-208 (1987).

[14] Goldwasser S., Sipser M.: Public coins vs private coins in interactive proofs systems. Proc. of STOC, 59-68 (1986).

[15] Hardy G.H., Wright E.M.: *An introduction to the theory of numbers.* Oxford University Press (1968).

[16] Kari J.: Reversibility of 2D cellular automata is undecidable. Physica D, 45, 379-385, (1990).

[17] Koebler J., Shoening U., Turan J.: *The Graph Isomorphism Problem - Its structural complexity.* Birkhauser ed, (1993).

[18] Martin O., Odlyzko A.M. and Wolfram S.: Algebraic properties of cellular automata. Comm. Math. Physics, 93, 219- (1984).

[19] Mac Williams F.J., Sloane N.J.: *The theory of error correcting codes.* North-Holland, Amsterdam (1977).

[20] Odlyzko A.M.: Discrete Logarithms in finite fields and their cryptographic significance. Bell Laboratories Internal Technical Memorandum, September, 27. (1983) (also in the Proceedings of CRYPTO 1985).

[21] Sutner K.: Computational complexity of finite cellular automata. Internal Note of the Stevens Institute of Technology, Hoboken, NJ 07030 USA (1989). (It will appear in Complex Systems).

[22] Wolfram S.: Computation theory of cellular automata. Comm. Math. Phys., 96 (1), 15-57 (1984).

[23] Wolfram S.: Twenty problems in the theory of cellular automata. Physica Scrypta, T9,170-183 (1985).

[24] Wolfram S.: Cryptography with cellular automata. Proc. of CRYPTO '85, Springer-Verlag (1985).

[25] Wolfram S.: Theory and application of cellular automata. World Scientific (1986).

[26] Wolfram S.: Random sequence generation by cellular automata. Advances in Appl. Math., 7, 123-169 (1986).

Parallel Pruning Decomposition (PDS) and Biconnected Components of Graphs

Eliezer Dekel

Jie Hu

IBM Israel
Science & Technology
Haifa, Israel 32000
edekel@vnet.ibm.com

Computer Science Program
University of Texas at Dallas
Richardson, TX 75083-0688
jhu@utdallas.edu

November 23, 1993

Abstract

We introduce pruning decomposition, a new method of graph decomposition which is very useful in developing a parallel algorithm for graph problems on EREW P-RAM. We present parallel algorithms that achieve the decomposition and an parallel algorithm for finding biconnected components of graphs based on the pruning decomposition. The complexity for both algorithms on EREW P-RAM is dominated by the spanning tree construction.

1 Introduction

Problem decomposition is a key to several algorithmic techniques. Methods of decomposition have the same basic feature. They break apart a complex structure of a given combinatorial object into smaller and simpler components. These simpler components are then processed in parallel to produce an efficient solution to the problem. Considerable effort in developing decomposition methods is reported in the literature. Many graph decompositions have been introduced for different type of graphs on different models of parallel machines. For example, ear decomposition is for biconnected graphs on CRCW or CREW P-RAM [MSV], tree contraction [ADKP] [NDP] [KR] and the centoid decomposition [CV] are for tree decompositions on EREW P-RAM.

In this paper we introduce *pruning decomposition*, a new graph decomposition, for general graphs on EREW P-RAM. Pruning decomposition partitions a graph of n vertices into $K \leq log n$ auxiliary graphs. Each of these auxiliary graphs consists of some simple structured components. The iterative natured pruning decomposition has the flavor of a general search technique in graphs (undirect and direct). It arranges the vertices of the graph by partitioning them into ordered sets of "chains". In this context we refer to the technique as *pruning decomposition search (PDS)*. In the modified version, the complexity of the PDS is dominated by the spanning tree construction, i.e., all the other operations can be performed in $O(logn)$ time using $n^2/logn$ processors on EREW.

The PDS can be applied to many graph problems. In the second part of this paper, we present a parallel algorithm for finding biconnected components of graphs which is a basic problem in graph theory and is used very often for many graph problems. The sequential algorithm to find the biconnected components is based on depth first search which has time complexity $O(m)$, where m is the number of edges in the given graph. The parallel algorithm on CRCW takes $O(logn)$ time using $m + n$ processors [TV]. This algorithm can be improved to $O(logn)$ time using $(m + n)\alpha \, (m, n)/logn$ processors, where α is the inverse Ackermann function [KR]. On CREW model of computations, this problem can be solved in $O(logn)$ time using $O(n\lceil n/log^2 n\rceil)$ processors [TC] or $O(n^2/p)$ time using p processors, where $p \leq n^2/logn$. Our algorithm runs in $O(logn)$ time using $n^2/logn$ processors on EREW based on the PDS. Other problems such as ear decomposition and st-numbering are also found $O(logn)$ time EREW algorithm based on the PDS [H]. For directed graphs, the PDS is applied to find minimum cutset of reducible graphs for program verifications and parallelized compiling [DH].

In the following, section 2 we introduce the pruning decomposition and describe the structure of the auxiliary graphs. We also present the parallel implementation of pruning decomposition. In section 3 we modify the pruning decomposition such that all the steps can be performed in $O(logn)$ time using $n^2/logn$ processors except the construction of the spanning tree. This modified version can be applied to problems for which the spanning tree can be efficiently found on EREW P-RAM and provides the possibility of obtaining more efficient ways for graph problems if better spanning tree construction method is developed (e.g., an efficient EREW connected component algorithm is recently presented by Chong and Lam [CL].) We apply the PDS for finding biconnected components of graphs in section 4, it can be extended to find articulation points, bridges and the 2-edge connected components.

2 Pruning Decomposition of Graphs

As we mentioned in the introduction, decomposition methods partition a problem so as to handle it in parallel. The pruning decomposition partitions the graph into *branches* (to be defined). It consists of four stages. We first give the description of these four stages, and then present our parallel decomposition algorithm.

Stage 1: Partition the spanning tree into sets of paths(chains)

Consider a rooted spanning tree T of an undirected graph G (Fig. 2.1). We define an *active chain* to be the longest subpath on the path from a leaf vertex to the root of T, starting from the leaf and including only vertices with degree

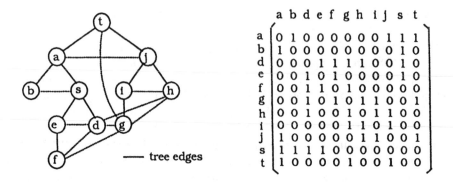

$$
\begin{array}{c}
\quad\; a\; b\; d\; e\; f\; g\; h\; i\; j\; s\; t \\
\begin{array}{c}
a \\ b \\ d \\ e \\ f \\ g \\ h \\ i \\ j \\ s \\ t
\end{array}
\left[
\begin{array}{ccccccccccc}
0 & 1 & 0 & 0 & 0 & 0 & 0 & 0 & 1 & 1 & 1 \\
1 & 0 & 0 & 0 & 0 & 0 & 0 & 0 & 0 & 1 & 0 \\
0 & 0 & 0 & 1 & 1 & 1 & 1 & 0 & 0 & 1 & 0 \\
0 & 0 & 1 & 0 & 1 & 0 & 0 & 0 & 0 & 1 & 0 \\
0 & 0 & 1 & 1 & 0 & 1 & 0 & 0 & 0 & 0 & 0 \\
0 & 0 & 1 & 0 & 1 & 0 & 1 & 1 & 0 & 0 & 1 \\
0 & 0 & 1 & 0 & 0 & 1 & 0 & 1 & 1 & 0 & 0 \\
0 & 0 & 0 & 0 & 0 & 1 & 1 & 0 & 1 & 0 & 0 \\
1 & 0 & 0 & 0 & 0 & 0 & 1 & 1 & 0 & 0 & 1 \\
1 & 1 & 1 & 1 & 0 & 0 & 0 & 0 & 0 & 0 & 0 \\
1 & 0 & 0 & 0 & 0 & 1 & 0 & 0 & 1 & 0 & 0
\end{array}
\right]
\end{array}
$$

Figure 2.1 Graph G, spanning tree T and matrix M

two or less (the root of the tree is included only if its degree is one or less). The *root of chain* is the end of the chain that is not the leaf. For a given tree T, we can find k active chains if there are k leaves. The first step of the decomposition is to move those active chains to an auxiliary graph, namely, "prune". The remaining tree has some new leaves generated, from which the new active chains can be found and move to another auxiliary graph. We say a vertex is *activated* (hence a *chain vertex*) when it is moved to an auxiliary graph. The iterative steps terminate once the root of the tree becomes active. Each auxiliary graph A_i (generated in i^{th} iteration) consists of some active chains. Obviously every vertex is activated exactly once. Actually this step is to partition T into several sets of chains.

It is instructive to follow the decomposition with an example. Figure 2.1 gives an undirected graph G, its spanning tree T with root t and its adjacency matrix M. Figure 2.2 shows the results of the first step: auxiliary graph A_1 with five chains, A_2 with two chains $a - s, j$ and A_3 with one chain t.

Lemma 2.1 *The total number of auxiliary graphs is no more than* $\log n$, *where n is the number of vertices in G.*

Proof: In an arbitrary tree T, if vertex v is the only child of vertex u then we say vertices u, v are in a same chain (this relation is transitive). A chain is *active* if it contains a leaf (it is exactly the definition of active chain in pruning decomposition) and is otherwise $non - active$. In the process of pruning decomposition, vertices on the same chain are on the same active chain in some iteration. Hence representing a non-active chain by one vertex will not change the number of iterations. Reduce T into T' by replacing each chain by a vertex so that every internal vertex of T' has at least two children, the number of leaves

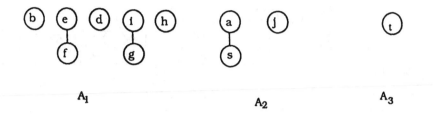

Figure 2.2 Auxiliary graphs of G by stage 1

(i.e., the active chains) in T' is at least half of the vertices. This implies the total number of iterations is bounded by $logn$.□

Stage 2: Construct frames of the branches

We augment the information in the auxiliary graphs by adding "super leaves" (to be defined) to the chains to build the *branch frames*. Let v be a chain vertex and u be a child of v in the spanning tree T. Vertex u was a root of an active chain in a previous iteration. We attach $\bar{u}$ to v as a *super leaf*. This super leaf represents the vertex set of the subtree of T rooted at u. The auxiliary graphs A_i's of our example together with the new super leaves are presented in Fig.2.3: super leaves $\bar{b}$, $\bar{e}$, $\bar{d}$, $\bar{i}$, $\bar{h}$ are added into A_2 and $\bar{a}$, $\bar{j}$ are added into A_3. In the following discussion we say that x is a member of $\bar{u}$, $x \in \bar{u}$ (x is represented by $\bar{u}$), if x is a descendant of u in T. Obviously there are no super leaf in A_1. Since the chains are disjoint, the super leaves are disjoint.

Lemma 2.2 *A vertex of G can be a super leaf at most once.*

Proof: A super leaf $\bar{u}$ is defined only the parent of u is a chain vertex. Since every vertex is activated (being a chain vertex) exactly once, a vertex can be a super leaf at most once. In fact, only the root of a chain can be a super leaf when its parent is activated.□

Stage 3: Complete branches by including representing edges

To complete our pruning decomposition, we need to include the edge information that is involved within each branch frame. We define the representing edges of each branch frame as following:

- If there is an edge (x, y) of G s.t. $x \in \bar{u}$, $y \in \bar{w}$ and $\bar{u}, \bar{w}$ are super leaves of the same chain, then $(\bar{u}, \bar{v})$ is the representing edge of (x, y);

- If there is an edge (x, v) of G s.t. $x \in \bar{u}$ and v is a chain vertex, then $(\bar{u}, v)$ is the representing edge of (x, v);

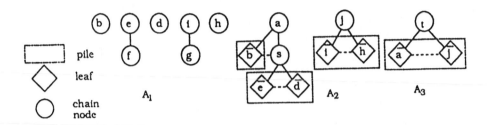

Figure 2.3 Auxiliary graphs of G

- If u, v are chain vertices and (u, v) is an edge of G, then (u, v) is the representing edge of itself.

Lemma 2.3 *Every edge of G is represented by some representing edge in auxiliary graphs exactly once.*

Proof: We prove by the type of edges based on the spanning tree T of G.

Case 1: edge (u, v) is a tree edge of T. Without loss of generality, let v be the child of u in T. If u, v are in the same active chain, then edge (u, v) is a chain edge which is represented by itself in the auxiliary graph where u is a chain vertex. If v is activated before u then v must be a root of a branch, hence $\bar{v}$ is a super leaf where u is a chain vertex. Edge (u, v) is represented only in the auxiliary graph in which u is a chain vertex.

Case 2: edge (u, v) is a forward edge of T. Without lost of generality, let v be the descendant of u in T. If u, v are in the same active chain, then similar to the case 1, (u, v) is represented by itself when u, v are chain vertices. If v is activated before u, then (u, v) is represented by $(u, \bar{w})$ where w is the child of u and w is an ancestor of v. Edge $(u, \bar{w})$ is in the auxiliary graph where u is a chain vertex. Hence (u, v) is represented only in the auxiliary graph where u is a chain vertex.

Case 3: edge (u, v) is a cross edge of T. We know that the representing edge for (u, v) is added only when u, v belong to the same branch. Let w be the lowest common ancestor of u and v in T. Consider the iteration when the representing edge of (u, v) is added in, if u is a chain vertex then (u, v) is represented by $(u, \bar{x})$ where x is the child of w and x is an ancestor of v. If v is a chain vertex the (u, v) is represented by $(\bar{y}, v)$ where y is the child of w and y is an ancestor of u. If none of u, v are chain vertices then (u, v) is represented by $(\bar{x}, \bar{y})$ where x, y are children of w, u is a descendant of x and v is a descendant of y. It is impossible for u and v be both chain vertices in a same chain since (u, v) is a cross edge. So (u, v) is represented only when the lowest common ancestor of u, v is activated.

Notice that every vertex of G is activated (i.e., to be a chain vertex) exactly once. Every edge of G is represented exactly once since it is represented only when certain vertex is an activated chain vertex.□

A branch frame with all the representing edges is referred to as a *branch*, and the root of the chain is referred to as the *root of branch*. From Lemma 2.3, we know that the PDS also partitions the edges of G.

Stage 4: Edge classification

Consider the relation of only the super leaves (excluding the chain vertices) in a branch, some of them are connected to each other. We define a *pile* to be a set of super leaves that are connected to each other in a branch. In the following, a pile is denoted by p_u if u is the vertex with the largest label in the pile (assume the vertices are labeled in postorder of the spanning tree T). Let b_w be a branch with root w, a pile p_u of b_w is *local* if there is no edge (x, y) of G s.t. x is in p_u and and y is not in b_w. Otherwise the pile is $non-local$. In our example in A_2 of Fig.2.3, p_d consisting of $\bar{e}$, $\bar{d}$ is a non-local pile because $(d, g) \in G$ has $d \in p_d$ but $g \in b_a$. Clearly, p_b is a local pile of b_a.

Furthermore, the edges of a branch can be partitioned into:

- Chain edges (the endpoints of the edges are chain vertices, e.g., (a, s) of A_2 in Fig.2.3);

- Edges between non-local piles and chain vertices (e.g., $(\bar{e}, s)$ of A_2 in Fig.2.3);

- Edges between local piles and chain vertices (e.g., $(a, \bar{b})$ of A_2 in Fig.2.3);

This classification concludes the pruning decomposition. We are now ready to describe our parallel implementation of pruning decomposition, Algorithm 2.1. In step 1 of PDS, we construct a spanning tree of the input graph. This can be done in $O(log^2 n)$ time using $n^2/logn$ processors on EREW P-RAM [NM], where the input is an adjacency matrix and the output of the tree is represented by adjacency lists. The postorder of T in step 2 can be computed in $O(logn)$ time using $n/logn$ processors on EREW [TV] [KR].

The stage 1 and stage 2 are performed by Step 4.1 of Algorithm 2.1. By Lemma 2.1 we know that step 4.1 repeats at most $logn$ times. In each iteration, 4.1.1 finds the longest possible chain from every leaf, which can be done in $O(logn)$ time using $n/logn$ processors by recursive doubling. In step 4.1.2 we attach the super leaves to the chain vertices. Therefore it can be performed on EREW in $O(logn)$ time using $n/logn$ processors..

Stage 3 is obtained by Step 2, which finds all the representing edges. This is done by using the information in adjacency matrix M. By the definition of

Procedure PDS

1. Construct a spanning tree T of G.
2. Relabel the vertices by the postorder of T and arrange M according to the new label;
3. { Initialization }
    3.1 $T_{org} \leftarrow T$;
    3.2 $i \leftarrow 0$;
4. { Construct the auxiliary graphs iteratively.}
    4.1 Repeat
        4.1.1 $i \leftarrow i+1$;
        4.1.2 find active chains and move them from T to A_i;
        4.1.3 include super leaves in each chain of A_i;
        Until T is empty;
    4.2 Include representing edges in A_i's;
    4.3 $K \leftarrow i$; { K auxiliary graphs formed }
5. { Find all the piles.}
    5.1 Construct $G_l = (V_l, E_l)$ where
        $V_l \leftarrow \{\bar{x} | \bar{x} \in A_i \text{ for some i}\}$;
        $E_l \leftarrow \{(\bar{x}, \bar{y}) | (\bar{x}, \bar{y}) \in A_i \text{ for some i}\}$;
    5.2 find a spanning forest F_l of G_l;
    { Every spanning tree of the super leaves indicates a pile.}
6. **for** i=1 to K **do** {Classification}
    for each branch b_w with root w of A_i **in parallel do**
    6.1 **for** each super leaf $\bar{u}$ of b_w **in parallel do**
        $\bar{u}$ is non-local if there is an edge (x, y) in G s.t. $x \in \bar{u}$ and
        y is not a descendant of w in T_{org}, otherwise $\bar{u}$ is local;
    6.2 **for** each pile of b_w **in parallel do**
        the pile is non-local if there is a u in the pile
        is non-local, otherwise the pile is local;
    6.3 classify the edges of the branch;
end;

Algorithm 2.1

the adjacency matrix, element $a(i,j)$ of M indicates there is an edge between
vertices i and j in G. We transfer the information of a column i of M into
spanning tree T of G, by assigning a value $R(i,j) = a(i,j)$ to vertex j of T. If
we update the values s.t. $R(i,j)=1$ if there is a descendant x of j has $R(i,x)=1$,
$R(i,j)=1$ indicates there is an edge between vertex i and some vertex in the
vertex set of subtree of T with root j, T_j. Using our example graph of Fig.2.1,
transferring and updating of column b in T is illustrated in Fig.2.4. This can be
done in $O(logn)$ time using $n/logn$ processors on EREW P-RAM using Euler
Tour [TV]. To obtain $R(i,j)$'s for all the columns we need $n^2/logn$ processors.
We keep these $R(i,j)$'s into a matrix M_1 in which $b(i,j)=R(i,j)=1$ indicates
there is an edge between vertex i and T_j. Based on M_1 we perform the similar
updating for each row and keep the results in M_2. Element $c(i,j)=1$ of M_2 means
there is an edge between vertices of T_i and T_j. Now we can add representing
edges into auxiliary graphs. Consider a branch of A_i, if i is a chain vertex, $\bar{j}$ is
a super leaf and $b(i,j)=1$ in M_1, then add representing edge $(i,\bar{j})$ because there
is an edge between vertex i and T_j in G. Similarly, if $\bar{i}, \bar{j}$ are super leaves and
$c(i,j)=1$ in M_2, add representing edge $(\bar{i}, \bar{j})$. So step 4.2 runs in $O(logn)$ time
using $n^2/logn$ processors and hence step 4 can be done in $O(log^2n)$ time using
$n^2/logn$ processors on EREW P-RAM.

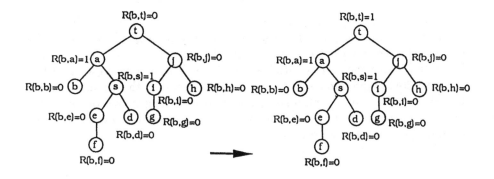

Figure 2.4 Computing $R(b,j)$ for column b in M on T

Stage 4 is completed by step 5 and step 6. By Lemma 2.2 and Lemma 2.3
we know that super leaves of all the auxiliary graphs are disjoint and so are the
representing edges. There is no conflict in generating a graph G_l that consists
only the super leaves of all the auxiliary graphs and the edges among the super
leaves in step 5.1 (Fig 2.5). In the spanning forest F_l of G_l that is found in step
5.2, every set of connected super leaves (or every spanning tree in F_l) is a pile.
For our example, F_l is the same as G_l, we can see that there are four piles p_b,

p_d, p_h and p_j (Fig.2.5). The complexity of step 5 is dominated by the spanning forest construction.

Step 6 has K$\geq$ *logn* iterations. In $i^t h$ iteration, 6.1 and 6.2 can be performed based on the matrices M_1 and M_2 Applying step 6 on our example, p_j and p_b are found to be local piles while the rest are non-local piles. This can be done in O(*logn*) time using $n^2/logn$ processors on EREW P-RAM. As long as we have the information of chain vertices, super leaves, piles (local or non-local) and F_l, the classification of edges can be performed in O(1) time using n^2 processors or in O(*logn*) time using $n^2/logn$ processors. Therefore step 6 can be done in O(*logn*) time using $n^2/logn$ processors on EREW.

Figure 2.5 Spanning forest of super leaves

In conclusion, the pruning decomposition of a given graph with n vertices can be achieved in O($log^2 n$) time using $n^2/logn$ processors on EREW P-RAM based on the tree and matrix data manipulations. This decomposition partitions an undirected graph into a set of auxiliary graphs, each of them consists of some branches which contains information about edges among the set of vertices the branch represents. This information can be utilized in obtaining efficient parallel solution for many graph problems.

3   Modified Pruning Decomposition

In the first step (step 4.1.2 of Algorithm 2.1) of the pruning decomposition described in previous section, we found the longest chain from each leaf. This is required for solutions of some graph problems [DH]. A more efficient version of pruning decomposition can be obtained when the chain does not have to be the longest possible. In this decomposition, the construction of the spanning tree dominates the complexity, while all the other steps can be done in O(*logn*) time using $n^2/logn$ processors on EREW P-RAM. Hence in graphs for which a spanning tree can be found efficiently on EREW P-RAM (such as planar graphs, serial parallel graphs, directed acyclic graphs, etc.), the pruning decomposition can be obtained much faster. In addition, any faster EREW spanning tree method will provide a better pruning decomposition with this modified version.

The modification is using tree contraction to partition the vertices into ordered chains, instead of finding longest chains from the leaves in each iteration.

There are various methods for tree contractions. Here we use the one introduced by K. Abrahamson et al [ADKP], which runs in $O(logn)$ time using $n/logn$ processors on EREW P-RAM. This algorithm works on binary trees. We can interpret the general tree T as a regular binary tree T' using Knuth's method [K] as follows: If v is a vertex of T with d children then the vertex set of T' includes $v^1, v^2, ..., v^{d+1}$. Vertex v^{i+1} is the left child of vertex v^i in T', for $1 \leq i \leq d$. Furthermore, if vertex w is the ith child of v in T the vertex w^1 is the right child of vertex v^i in T' (see Fig. 3.1).

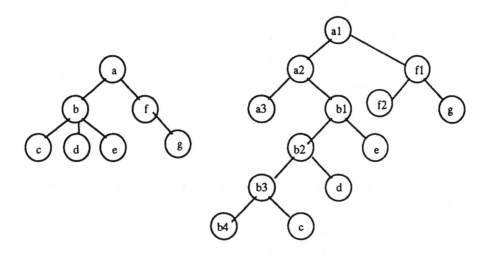

Figure 3.1 Interpretation of trees as binary trees

The tree contraction based on the binary tree is briefly presented by the procedure CONTRACT.

We modify the function bypass(w) by letting $w - v$ to be a chain indicated by v, which can be considered as a vertex chain (can be implemented as a link list). Initially every vertex is a chain with one vertex. When calling bypass(w), the chain indicated by w is concatenated with the chain indicated by v and the new chain is indicated by v. The function prune(v) moves the chain indicated by v to A_i instead of simply removing. By this modification, in each iteration a set of chains are moved to an auxiliary graph and all the descendants of the vertices of the chain are already "pruned" in previous iterations. This is exactly the properties we need for the pruning decomposition. Clearly the descendants

Procedure CONTRACT(T)

```
for each leaf v in parallel do
    index(v) ← left_to_right leaf index of v
repeat ⌈logN − 1⌉ times
    for each leaf v in parallel do
        w ← parent(v)
        if index(v) is odd and w ≠ root
        then if v is a left child
            then prune(v);
                bypass(w);
    for each leaf v in parallel do
        w ← parent(v)
        if index(v) is odd and w ≠ root
        then if v is a right child
            then prune(v)
                bypass(w)
    index(v) ← index(v)/2
end;
```

of v^i in T' are the descendants of v in T. If we let v^1 of T' represents v and skip all the $v^2, ..., v^d$ in each chain, the properties for the pruning decomposition still hold. Hence we can use the tree contraction to partition the vertices of G into ordered vertex chains. To expand pruned nodes into vertex chains takes at most $logn$ times using $n/logn$ processors on EREW since those chains are disjoint.

The algorithm of this modified pruning decomposition is derived from the procedure PDS (Algorithm 2.1). The only change is in the construction of auxiliary graphs (step 4 of Algorithm 2.1). Replacing step 4 of Algorithm 2.1 by the modified tree contraction as described previously gives the implementation of the modified pruning decomposition.

The algorithm of modified pruning decomposition is dominated by spanning tree construction (step 1 and step 5 of Algorithm 2.1). All the other steps can be achieved in $O(logn)$ time using $n^2/logn$ processors. For the graphs whose spanning tree can be found efficiently, the pruning decomposition can be obtained faster. When a graph is a tree, no spanning tree needs to be constructed and no representing edges need to be updated, the pruning decomposition can be done in $O(logn)$ time using $O(n/logn)$ processors on EREW P-RAM.

4 Biconnected components of graphs

We first introduce the idea of how our algorithm finds the biconnected components based on the pruning decomposition and then proceed to discuss the details of the parallel implementation.

Let BC be the set of biconnected components and AP be the set of articulation points of graph G. A $pseudo - graph$ G_{ps} can be derived from G in the following way:

- Every biconnected component $B_i \in BC$ is a node in G_{ps};

- Two biconnected components B_i and B_j are "connected" by a vertex u if $u \in B_i$, $u \in B_j$ and $u \in AP$.

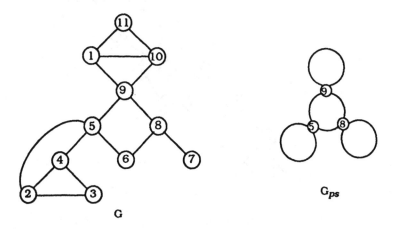

G

Figure 4.1 Pseudo-graph of G

Figure 4.1 shows an example of pseudo-graph G_{ps} of G. Notice that G_{ps} has a structure similar to a tree. A biconnected component B is a $pendant$ if it contains only one articulation point (we define this articulation point the $anchor$ of the biconnected component). There must be some pendants in any G_{ps} according to the tree like structure. If we remove the pendants (except the articulation points which are contained in the non-pendants also), some new pendants will be generated in the remaining part of G_{ps}. The strategy that keeps finding and removing the pendants makes every B a pendant exactly once and leads us to a method which can be modified into a parallel algorithm for finding the biconnected components of a graph. Clearly every biconnected component has exact one anchor, except the last remained biconnected component.

We are using the results of the pruning decomposition to perform this strategy. For the purpose of finding biconnected components, we need to define some more terms related to the auxiliary graphs of pruning decomposition for further presentation. In a branch b_w, a chain vertex u is $non - local$ if u can reach out of b_w (i.e., a vertex v that is not a descendant of w in T) without passing the parent of u. Otherwise u is *local*. A super leaf $\bar{u}$ is *attached* to a chain vertex x if there is an edge $(x,\bar{u})$ in the auxiliary graph. Similarly, a chain vertex x is attached by a pile if there is a super leaf attaches to x belongs to the pile. Let $max(p_u)$ ($min(p_u)$) be the largest (smallest) chain vertex attached by p_u. A chain vertex x is *internal* to p_u if $max(p_u) > x > min(p_u)$.

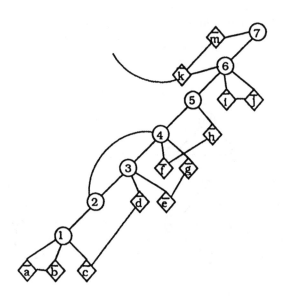

Figure 4.2 Finding articulation points in a branch

We proceed to show that the articulation points can be found locally in the branches of the auxiliary graphs from the pruning decomposition of G.

Consider a branch of an auxiliary graph (Fig.4.2). There are some local and non-local piles attach to the chain vertices. Assume the vertices are labeled in postorder of T. We have the following observations:

1. If a chain vertex u is non-local and chain vertex $x > u$, then x is also non-local (in Fig.4.2, vertex 7 is nonlocal because vertex 6 is nonlocal).

2. If a local pile p_u attaches only to one chain vertex x, i.e., $max(p_u) = x = min(p_u)$, x is an articulation point (called *type I* articulation point) that separates the vertices belong to p_u from other vertices (in Fig.4.2, vertex 1,6 are type I articulation points).

3. If a local chain vertex (or the smallest non-local chain vertex but not the smallest chain vertex) x is not internal to any of local piles, than x is an

articulation point (called *type II* articulation point) that separates its parent from its children (in Fig.4.2, vertices 4,5 and 6 are type II articulation points).

Lemma 4.1 *A vertex u is an articulation point of G iff u is identified in a branch as an articulation point of type I or type II.*

Proof: "$\longrightarrow$": If u is an articulation point of G, u must separate its parent v from one of its children x. In the branch where u is a chain vertex, we have the following cases:

Case 1: Both u and x are chain vertices of the branch. Because u separates v and x, there is no pile (or forward edge) that attaches chain vertices larger than u and smaller than u, i.e., u is not internal to any local pile (or forward edge). It is identified as a type II articulation point.

Case 2: u is a chain vertex but $\bar{x}$ is a super leaf which belongs to a local pile p_y. If p_y attaches only to u then u is identified as a type I articulation point. If p_y attaches to other chain vertices, since v cannot reach x without passing u, u must be either $max(p_y)$ or $min(p_y)$. Therefore u is not internal to any other local pile. It is identified as an articulation point of type II. *Case 3:* u is the root of the branch. If x is a chain vertex, then u is either a local or the smallest non-local chain vertex. It cannot be internal to any pile or forward edge since it is the root of the branch. Hence it is identified as an articulation point of type II. If x is a super leaf, it belongs to some local pile p_y. Similar to case 1 and 2, u is identified as an articulation point of type I or type II.
"$\longleftarrow$": Trivial.$\square$

Consider a branch without any of the local piles that attach to only one chain vertex. If there is no type II articulation point then all the chain vertices belong to a same biconnected component, because no chain vertices can separate the chain. Therefore all the vertices of super leaves belong to the same biconnected component. If there are some type II articulation points, let x be the largest type II articulation point. Because x separates the upper part of the branch from the lower part of the branch and there are no type II articulation points greater than x, all the chain vertices that are not smaller than x and all piles attach only to those chain vertices belong to a same biconnected component. All the other vertices of the branch are of other biconnected components. If we remove those vertices of other biconnected component, $\bar{w}$ will represent vertices of a same biconnected component in some auxiliary graph later (defined to be *purified* super leaf), such as $\bar{i,j}$, 7 and 8 will be represented by purified super leaf $\bar{8}$. Since the root of the branch is never removed and all the removed vertices are either local chain vertices or of local piles, the structure of all the branches will not change.

We proceed to discuss how to identify biconnected components locally in a branch.

Given a branch of with all the super leaves "purified". We have:

1. If u is a type I articulation point and p_x is a local pile that attaches only to u, then vertices of p_x and u belong to a same biconnected component (such as $\bar{a}$, $\bar{b}$ and vertex 1 of Fig.4.2). According to the property of purified super leaf, there are no other vertices belong to this biconnected component. It is a biconnected component of *type I* and u is its anchor.

2. If u be a type II articulation point and v be the greatest chain vertex that is smaller than u and is not internal to any of the local piles (such as 4 and 1 in Fig.4.2). All the chain vertices between u and v (including u and v) and local piles attached to those vertices are of a same biconnected component (such as 1, 2, 3, 4, $\bar{c}$, $\bar{d}$, $\bar{e}$ and $\bar{g}$ in Fig.4.2). By the property of purified super leaf, it is a biconnected component of *type II* and u is its anchor.

Lemma 4.2 *A subgraph $B = (V_B, E_B)$ is a biconnected component of G iff V_B is identified as a biconnected component in some branch.*

Proof: "$\longrightarrow$:" Let B be a biconnected component of G and u be its anchor. By the structure of G_{ps}, the anchor of B has the greatest preorder among the vertices of B. Hence every vertex of B is either a chain vertex or belongs to some super leaf in the branch where u is a chain vertex. Here we have two cases:

Case 1: u is the only vertex of B that is a chain vertex of the branch, while all the other vertices of B are represented by some super leaves. Since B is a connected component, those super leaves belong to pile p_y. Since B is biconnected maximally, p_y is local. Obviously p_y attaches only to u and u is a type I articulation point. By the discussion above, all the vertices of B belong to $p_y \cup \{u\}$ and $p_y \cup \{u\}$ contains only vertices of B. It is identified as a type I biconnected component in this branch.

Case 2: There are some vertices other than u of B are chain vertices of the branch. Let v be the smallest vertex of B that is also a chain vertex. Every chain vertex between v and u is internal to some local piles or forward edges, otherwise v and u can be separated by that vertex. Chain vertex v is not internal to any local pile or forward edge because (1) If v is the smallest vertex of the branch then it is not internal to any local piles, (2) if v has a child x that is also a chain vertex then v is not internal to any local piles, otherwise x belongs to B and $x < v$, which contradicts the assumption. Therefore v is the greatest chain vertex that is smaller than u and is not internal to any local piles. Every vertex of B is either a chain vertex between u and v or belong to some local pile that attaches to those vertices. By the property of purified super leaf, B is identified as a type II biconnected component in this branch.
"$\longleftarrow$": Trivial.□

Since the articulation points and biconnected components can be found locally in branches, we are ready to present our method to find articulation points and biconnected components of graph G based on the pruning decomposition. The parallel implementation of our method is shown in Algorithm BI_COMPONENTS (Algorithm 4.1).

Algorithm BI_COMPONENT

{*Input*: Results of pruning decomposition;
Output: Set of articulation points AC and set of biconnected component BC.}
1. Initialization
    $AP \leftarrow \emptyset;\ BC \leftarrow \emptyset;$
2. Find $max(p_u)$ and $min(p_u)$ for each local pile p_u;
3. Mark chain vertices that are internal to some local piles or forward edge;
4. Identify "non_local" chain vertices;
5. "Purifying" super leaves;
6. Find articulation points and biconnected components
    For each branch in parallel do
    6.1 Identify type I articulation points and biconnected components
        For each local pile p_u with $max(p_u) = min(p_u)$ in parallel do
            6.1.1 $AP \leftarrow AP \cup \{\max(p_u)\};$
            6.1.2 $S \leftarrow p_u \cup \{\max(p_u)\};$
            6.1.3 $BC \leftarrow BC \cup \{S\};$
    6.2 Identify type II articulation points and add them into AP;
    6.3 Identify type II biconnected components
        For each type II articulation point u in parallel do
            6.3.1 $v \leftarrow max\{x | x < u$ is a non_internal chain vertex$\};$
            6.3.2 $A \leftarrow \{x | v \geq x \geq u$ and x is a chain vertex$\};$
            6.3.3 $B \leftarrow \{x | x$ belongs to a pile that attaches only to a vertex of $A\};$
            6.3.4 $C \leftarrow A \cup B;$
            6.3.5 $BC \leftarrow BC \cup C;$
end;

Algorithm 4.1

The inputs of the BI_COMPONENTS, Algorithm 4.1, are the results of the pruning decomposition. Initially, the set of articulation points AP and the set of biconnected components BC are empty. The algorithm works locally on each branch. The iterative strategy of pseudo graph G_{ps} can be performed without iterations since the pruning decomposition gives the disjoint branches.

Step 2 through step 5 can be done based on the matrices M, M_1 and M_2 from the pruning decomposition. The operations involve OR, maximum finding, minimum finding and deleting on rows or on columns. These can be performed in $O(logn)$ time using $n^2/logn$ processors on EREW P-RAM.

Step 6 identifies the articulation points and biconnected components locally on each branch. In step 6.1, the type I articulation points and biconnected components can be found in $O(1)$ time using n processors. Step 6 can obtains all the type II articulation points in $O(1)$ time using n processors. In step 6.3.1, the greatest non-internal vertex can be found by partial sum which can be achieved in $O(logn_b)$ time using $n_b/logn_b$ processors, where n_b is the number of chain vertices of the branch. Hence it can be achieved in $O(logn)$ time using $n/logn$ processors for all the branches. The rest of step 6.3 can be done in $O(logn)$ time using $n^2/logn$ processors on EREW P-RAM. Therefore the whole algorithm can be done in $O(logn)$ time using $n^2/logn$ processors on EREW P-RAM.

Theorem 4.1 *Algorithm BLCOMPONENTS (Algorithm 4.1) finds all the bi-connected components of G in O(logn) time using $n^2/logn$ processors on EREW P-RAM.*

Proof: Follows by Lemma 4.1, Lemma 4.2 and the complexity analysis of the algorithm.□

As an extension of this result, bridges and 2-edge connected components of an undirected graph G can be found with the same complexity.

Notice the facts that every bridge of G must be a tree edge of T and the end point of a bridge must be either an articulation point or a vertex of degree one. So we can focus on just tree edges, articulation points, leaf and root of T. If both the end points of a bridge are activated in the same iteration, i.e., they are adjacent chain vertices then they cannot reach each other without passing the bridge. If the endpoints are activated in different iteration, then the bridge is represented by an only edge between a local pile containing only one super leaf and a chain vertex. Therefore according to the algorithm BLCOMPONENTS (Algorithm 4.1), if a chain vertex u is identified as a type I articulation point by step 6.1, let the super leaf attaches to u is $vbar$ (i.e., v is a child of u in T), then (u, v) is a bridge when v is a type II articulation point in the branch with root v. If a chain vertex u is a type II articulation point and in step 5.3.1 v is the only child of u in T, then (u, v) is a bridge when v is also a type II articulation point and no pile attaches to both u and v. Obviously adding these operations into the algorithm will not change the complexity. Moreover, if we delete all the bridges from the graph, every connected component will be a 2-edge connected component. So all the bridges and 2-edge connected components of G can be found in $O(logn)$ time using $n^2/logn$ PE's on EREW P-RAM based on the pruning decomposition.

We have shown that all the articulation points, bridges, biconnected components and 2-edge connected components can be found efficiently on EREW P-RAM based on the pruning decomposition. Since the complexity of pruning decomposition is dominated by the spanning tree construction, our algorithm is particularly good for graphs where the spanning tree can be constructed efficiently on EREW P-RAM.

References

[ADKP] K. Abrahamson, N. Dadoun, D. Kirkpatrick, and T. Prsytycka, "A simple Parallel Tree Construction Algorithm," *Journal of Algorithms*, 10, pp.287-302, 1989.

[CL] K. W. Chong and T. W. Lam, "Finding Connected Components in O($lognloglogn$) Time on the EREW PRAM," *SODA*, 1993.

[CV] R. Cole and U. Vishkin, "Optimal Parallel Algorithms for Expression Tree Evaluation and List Ranking," *AWOC* pp.91-100, 1988.

[DH] E. Dekel and J. Hu, "A Parallel Algorithm for Finding Minimum Cutsets in Reduceble graphs," *Journal of Parallel and Distributed Processing*, to appear.

[NDP] S. Ntafos, E. Dekel and S. Peng, "Compression Trees and Their Applications," *Proc. ICPP*, pp.132-139, 1987.

[H] J. Hu, "Parallel Algorithms for Distributed Systems and Software Engineering," *Ph. D. Dissertation, University of Texas at Dallas*, 1992.

[K] D. Knuth, "The Art of Computer Programming. Vol. 1. Fundamental Algorithms," Addison-Wesley, Reading, MA, 1968.

[KR] R. M. Karp and V. Ramachandran, "A Survey of Parallel Algorithms for Shared-Memory Machines," *Handbook of Theoretical Computer Science*, Cambridge, MIT Press, 1990.

[MSV] Y. Maon, B. Schieber and U. Vishkin, "Parallel Ear Decomposition Search and st-Numbering in Graphs," *Theo. Com. Sci.*, Vol.47, pp.277-298, 1986.

[NM] D. Nath and S.N. Maheshwari, "Parallel Algorithms for the Connected Components and Minimal Spanning Tree Problems," *Inf. Proc. Lett.*, Vol 14, No.1, pp.7-11, 1982.

[TC] Y. H. Tsin and F. Y. Chin, "Efficient Parallel Algorithms for a Class of Graph Theoretic Problems," *SIAM J. Comput.* Vol.13, No.3, pp.580-599, 1984.

[TV] R. E. Tarjan and U. Vishkin, "An Efficient Parallel Biconnectivity Algorithm," *SIAM J.Comput.*, Vol.14, No.4, 862-874, 1985.

A Non-Interactive Electronic Cash System [*]

(Extended Abstract)

Giovanni Di Crescenzo

Dipartimento di Informatica ed Applicazioni,
Università di Salerno, 84081 Baronissi (SA), Italy

Abstract. We give the first electronic cash system whose transactions require no interaction between the parties. Its security is based on the computational intractability of factoring Blum integers.
We also give a non-interactive perfect zero-knowledge proof system of membership to the language of Blum integers. We use this proof system in our electronic cash system as a tool for proving the correctness of some computations modulo Blum integers without revealing the factorization of such integers.

1 Introduction

Electronic cash systems. Electronic cash is one of the most important applications of modern complexity-based cryptography. While real-life cash bases its security on the physical difficulty of reproducing coins and bills, an electronic cash system must guarantee its security by mathematical arguments. In this situation cryptography plays a very important role.

Several electronic cash systems have been proposed (see [9, 10, 11, 15, 17, 19, 25]), all differing in security, privacy and efficiency features (see [15, 25]).

In this paper we concentrate our attention on the communication complexity of electronic cash systems. The problem of reducing the communication needed for cryptographic protocols has recently emerged as a major line of research in the area of cryptography. [1] were the first to investigate round complexity for secure function evaluation and exhibited a non-cryptographic method that saves a logarithmic number of rounds. In [2] it is shown how any function can be securely computed using only a constant number of rounds of interactions, under the assumption that one-way functions exist. The problem we consider is that of minimizing the number of rounds of communications between the parties involved in the protocols of an electronic cash system. Many systems proposed are quite efficient from a communication point of view (see [10, 11, 15, 19]) and have constant-round protocols for each transaction.

The main result of this paper is a non-interactive electronic cash system. That is, in our system each transaction is performed in the following way: a party sends a message to the other party involved in the operation and possibly

---

[*] Work supported by MURST and CNR.

writes a message on a public file; the other party accepts or rejects the message received. We underline that our system is the first with this property. Moreover, our electronic cash system enjoys the following properties: its security does not depend on any physical assumption; actually, it is based on the computational intractability of factoring Blum integers. Also, it is not possible to spend a coin not owned, to spend twice the same coin, or to trace the users involved in spending protocols. However, it is possible to spend a coin in an off-line manner, that is without the help of the Bank, to divide a coin in smaller pieces and to spend any desired fraction of a coin.

Zero Knowledge. A Zero-Knowledge proof is a special kind of proof that allows an all-powerful prover to convince a poly-bounded verifier of the veridicity of a certain statement without revealing any additional information. This concept has been introduced in [22]. A proof is *computational* or *perfect* zero knowledge if it does not reveal any information respectively to a poly-bounded or unlimited-power verifier. Computational zero knowledge proofs have been given for all languages in NP first ([21]) and then for all languages having a proof system ([4, 23]). Perfect zero-knowledge proofs are very unlikely to exist for all languages in NP because of their complexity consequences (see [18, 8]) and have been given only for a handful of number-theoretic and random self-reducible languages in NP (see [22, 21, 20, 26, 5]).

In the original model of [22], prover and verifier could use private coins and the proof was a result of an interaction between the two parties. In [7, 6] a new model for zero-knowledge has been put forward: the shared-string model. Here, the prover and the verifier share a random string and the communication is monodirectional. That is, the prover sends the (non-interactive) proof to the verifier that, after some computations, decides whether to accept or not. Although the shared-string model seems very restricted, all NP languages have computational zero-knowledge proofs in this model [6, 14, 16, 3]. Instead, non-interactive perfect zero-knowledge proofs have been given in [6, 12, 13], only for a class of languages based on quadratic residuosity.

In this paper we give a non-interactive perfect zero-knowledge proof for the language of Blum integers. In our proof system, the prover's algorithm can be performed in polynomial-time provided that a suitable witness (e.g., the factorization of the modulus) is given as an additional input. The proof is used in our electronic cash system as a tool for proving the correctness of some computations modulo Blum integers without revealing the factorization of such integers.

Organization of the paper.

In Section 2 we review some number-theoretic results about Blum integers.

In Section 3 we review the definition of [6] of non-interactive perfect zero-knowledge proof systems.

In Section 4 we present a non-interactive perfect zero-knowledge proof system of membership to the language of Blum integers.

In Section 5 we introduce the concept of non-interactive electronic cash system.

In Section 6 we present our non-interactive electronic cash system. We describe the protocols for the following transactions: starting up the system, opening an account, withdrawing a coin, spending a coin, spending a fraction of a coin and depositing a coin.

Finally, in Section 7 we give a sketch of proof of security of our non-interactive electronic cash system.

2 Background and Notations

2.1 Notations

We identify a binary string σ with the integer x whose binary representation is σ. If σ and τ are binary strings, we denote their concatenation by either $\sigma \circ \tau$ or $\sigma\tau$. By the expression $|x|$ we denote the length of x if x is a string. By the expression $\vec{w}$ we denote the k−tuple $(w_1, \ldots, w_k)$ of numbers.

2.2 Number Theory

We refer the reader to [24] and [6] for the definitions of quadratic residues, Jacobi symbol and Regular(s) integers, and for discussions about the computational intractability of computing square roots modulo a composite integer.

We define the *quadratic residuosity predicate* as $Q_x(y) = 0$ if y is a quadratic residue modulo x and 1 otherwise. Moreover, we let Z_x^{+1} and Z_x^{-1} denote, respectively, the sets of elements of Z_x^* with Jacobi symbol $+1$ and -1 and $QR_x = \{y \in Z_x^{+1} | Q_x(y) = 0\}$, $NQR_x = \{y \in Z_x^{+1} | Q_x(y) = 1\}$.

In this paper we will be mainly concerned with the special moduli called Blum integers.

Definition 1. An integer x is a Blum integer, in symbols $x \in \mathrm{BL}$, if and only if $x = p^{k_1} q^{k_2}$, where p and q are different primes both $\equiv 3 \bmod 4$ and k_1 and k_2 are odd integers.

From Euler's criterion it follows that, if x is a Blum integer, $-1 \bmod x$ is a quadratic non residue with Jacobi symbol $+1$. Moreover we have the following fact.

Fact 1. On input a Blum integer x, it is easy to generate a random quadratic non residue in Z_x^{+1}: randomly select $r \in Z_x^*$ and output $-r^2 \bmod x$.

The following lemmas prove that the Blum integers enjoy the elegant property that each quadratic residue has a square root which is itself a quadratic residue. Thus each quadratic residue modulo a Blum integer has also a fourth root.

Lemma 2. Let x be a Blum integer. Every quadratic residue modulo x has at least one square root which is itself a quadratic residue modulo x.

On the other hand, if x is a product of two prime powers, but not a Blum integer, then the above lemma does not hold.

Lemma 3. Let $x = p^{k_1} q^{k_2}$, where $p \equiv 1 \bmod 4$. Then, at least one half of the quadratic residues has no square root which is itself a quadratic residue modulo x.

The following characterization of Blum integers will be used to obtain a non-interactive perfect zero-knowledge proof system for the set of Blum integers.

Fact 2. An integer x is a Blum integer if and only if $x \in Regular(2)$, $-1 \bmod x \in NQR_x$, and for each $w \in QR_x$ there exists an r such that $r^4 \equiv w \bmod x$.

3   Non-Interactive Perfect Zero Knowledge

Let us now review the definition of Non-Interactive Perfect Zero Knowledge of [6] (we refer the reader to the original paper for motivations and discussion of the definition).

We denote by L the language in question and by x an instance to it. Let P a probabilistic Turing machine and V a deterministic Turing machine that runs in time polynomial in the length of its first input.

Definition 4. We say that (P, V) is a Non-Interactive Perfect Zero-Knowledge Proof System for the language L if there exists a positive constant c such that:

1. *Completeness.* $\forall x \in L$, $|x| = n$ and for all sufficiently large n,

$$\mathbf{Pr}(\sigma \leftarrow \{0,1\}^{n^c}; Proof \leftarrow P(\sigma, x): V(\sigma, x, Proof) = 1) > 1 - 2^{-n}.$$

2. *Soundness.* For all probabilistic algorithms $Adversary$ outputting pairs $(x, Proof)$, where $x \notin L$, $|x| = n$, and all sufficiently large n,

$$\mathbf{Pr}(\sigma \leftarrow \{0,1\}^{n^c}; (x, Proof) \leftarrow Adversary(\sigma): V(\sigma, x, Proof) = 1) < 2^{-n}.$$

3. *Perfect Zero Knowledge.* There exists an efficient simulator algorithm S such that $\forall x \in L$, $|x| = n$, and for all sufficiently large n, the two probability spaces $S(x)$ and $View_V(x)$ are equal, where by $View_V(x)$ we denote the probability space

$$View_V(x) = \{\sigma \leftarrow \{0,1\}^{n^c}; Proof \leftarrow P(\sigma, x) : (\sigma, Proof)\}.$$

We say that (P, V) is a Non-Interactive Proof System for the language L if completeness and soundness are satisfied. We call the "common" random string σ, input to both P and V, the *reference string*.

4 Non-Interactive Perfect Zero Knowledge for BL

In this section we show how to exploit the number-theoretic properties of Blum integers to obtain a non-interactive perfect zero-knowledge proof system (A,B) for the language BL.

The prover A wants to convince the polynomial-time verifier B that a natural number x is a Blum integer without giving away any information that B was not able to compute alone before. B cannot compute by himself whether $x \in$ BL, since the fastest way known for deciding this consists of factoring x, and for this problem no efficient algorithm is known. Moreover, the proof is non-interactive (A gives only one message to B), and perfect zero-knowledge (B does not gain any additional information even if not restricted to run in polynomial time).

In constructing our proof system, we essentially use Fact 2 in which a Blum integer is characterized as a $Regular(2)$ integer with the additional properties that $-1 \bmod x$ is a quadratic non residue with Jacobi symbol $+1$ and that each quadratic residue modulo x has a square root which is itself a quadratic residue.

Let us informally describe the protocol. A and B share a string τ of length $60n^2$. This string is split into $\rho_1 \circ \cdots \circ \rho_{50n} \circ \sigma_1 \circ \cdots \circ \sigma_{10n}$, where the ρ_i and σ_i have all length n. Let $x \in$ BL. First, A proves that $-1 \bmod x$ is a quadratic non residue modulo x using the protocol of [6]. While proving this, he also implicitly proves that Z_x^{+1} is partitioned by the relation $\sim_x$ into 2 equivalence classes, and this means that $x \in Regular(2)$. Then, for each $\sigma_i \in Z_x^{+1}$, A gives a random fourth root of σ_i or $-\sigma_i \bmod x$ according to whether σ_i is a quadratic residue or not. This can obviously be done because of Lemma 2 and the fact that $-1 \bmod x$ is a quadratic non residue. B verifies that $-1 \bmod x$ is a quadratic non residue with Jacobi symbol $+1$, that Z_x^{+1} is partitioned by the relation $\approx_x$ into 2 equivalence classes and that he has received a fourth root of σ_i or $-\sigma_i \bmod x$. If this is the case B accepts.

A formal description of (A,B) can be found in Figure 1.

Theorem 5. (A,B) is a Non-Interactive Proof System for BL.

Proof's sketch: First of all B runs in polynomial time. In fact, the Jacobi symbol can be computed in polynomial time, steps B.2.1, B.2.2 and B.2.4 are trivial, and step B.2.3 can be efficiently performed using a primality test (see [6] for details).

Let us now prove that completeness and soundness are satisfied.

Completeness. Assume $x \in$ BL. Then it is easy to see that steps B.2.1, B.2.2 and B.2.3 are passed with very high probability. Moreover, step B.2.5 is always passed. In fact, as $x \in Regular(2)$, there are exactly 2 $\sim_x$ equivalence classes in Z_x^{+1}. That is, either ρ_i is a quadratic residue modulo x or ρ_i is in the same equivalence class as $-1 \bmod x$, in which case $-\rho_i \bmod x$ is a quadratic residue. Thus A can always give a square root of ρ_i or $-\rho_i \bmod x$. Let us see that verification step B.3 is always passed. If σ_i is a quadratic residue modulo x, then by Lemma 2, it has a square root which is itself a quadratic residue. Instead, if

---

Input to A and B:
- A $60n^2$-bit random string τ.
- $x \in BL$, where $|x| = n$.

  (Set $\tau = \rho_1 \circ \cdots \circ \rho_{50n} \circ \sigma_1 \circ \cdots \circ \sigma_{10n}$, where ρ_i and σ_i have length n for each i.)

Instructions for A.

A.1 Set Proof = empty string.

A.2 (*Prove that x is a Regular(2) integer and -1 is a quadratic non residue modulo x.*)

  For $i = 1, \ldots, 50n$,

    if $\rho_i \in Z_x^{+1}$ then

      if $\rho_i \in NQR_x$ then randomly choose u_i such that $u_i^2 \equiv -\rho_i \bmod x$;

      if $\rho_i \in QR_x$ then randomly choose u_i such that $u_i^2 \equiv \rho_i \bmod x$;

    else set $u_i \leftarrow 0$;

    append (ρ_i, u_i) to Proof.

A.3 (*Prove that each quadratic residue modulo x has a fourth root modulo x.*)

  For $i = 1, \ldots, 10n$,

    if $\sigma_i \in Z_x^{+1}$ then

      if $\sigma_i \in NQR_x$ then randomly choose v_i such that $v_i^4 \equiv -\sigma_i \bmod x$;

      if $\sigma_i \in QR_x$ then randomly choose v_i such that $v_i^4 \equiv \sigma_i \bmod x$;

    else set $v_i \leftarrow 0$;

    append (σ_i, v_i) to Proof.

A.4 Send Proof.

Input to B:
- The string Proof $= ((\rho_1, u_1), \ldots, (\rho_{50n}, u_{50n}), (\sigma_1, v_1), \ldots, (\sigma_{10n}, v_{10n}))$ sent by A.

Instructions for B.

B.1 If $\rho_i \in Z_x^{+1}$ for less than $5n$ indices i, or $\sigma_i \in Z_x^{+1}$ for less than n indices i, then HALT and ACCEPT.

B.2 (*Verify that x is a Regular(2) integer and -1 is a quadratic non residue modulo x.*)

  B.2.1 Verify that x is odd.

  B.2.2 Verify that x is not a perfect square over the integers.

  B.2.3 Verify that x is not a prime power.

  B.2.4 Verify that $-1 \bmod x$ has Jacobi symbol $+1$.

  B.2.5 For $i = 1, \ldots, 50n$,

    if $\rho_i \in Z_x^{+1}$ then

      verify that $u_i^2 \equiv \rho_i \bmod x$ or $u_i^2 \equiv -\rho_i \bmod x$.

B.3 (*Verify that each quadratic residue modulo x has a fourth root modulo x.*)

  For $i = 1, \ldots, 10n$,

    if $\sigma_i \in Z_x^{+1}$ then

      verify that $v_i^4 \equiv \sigma_i \bmod x$ or $v_i^4 \equiv -\sigma_i \bmod x$.

If all verifications are successful then ACCEPT else REJECT.

Fig. 1. The proof system (A,B) for BL.

σ_i is a quadratic non residue modulo x, then $-\sigma_i \bmod x$ is a quadratic residue. Thus A can always give a random fourth root of σ_i or $-\sigma_i \bmod x$.

Soundness. Suppose that $x \notin$ BL. Then three cases are possible, in each of them B rejects with very high probability. First, it may be that $x = p^{k_1} q^{k_2}$, with $p \equiv q \equiv 3 \bmod 4$, but k_1 (or k_2) is even; in this case $(-1|x) = -1$ (and B can compute the Jacobi symbol in polynomial time). Second, it may be that $p \equiv 1 \bmod 4$, and then by Lemma 3, A has a negligible probability of giving the fourth roots of all σ_i's. Third, it may be that $x \notin Regular(2)$. Then, either x is a prime power (and this can be detected by B in step B.4) or x has more than two prime factors (but then A can prove that Z_x^{+1} is partitioned by the relation $\sim_x$ into 2 equivalence classes only with negligible probability). $\qquad\Box$

To prove the perfect zero-knowledge property, we describe an efficient simulator M such that, for each $x \in$ BL, the probability space $M(x)$ is equal to $View_B(x)$.

Let us informally describe the program M. It works in two phases. In the first phase M generates the random integers $\rho_1, \ldots, \rho_{50n}$ and simulates the proof that $-1 \bmod x$ is a quadratic non residue modulo x and that x is a Regular(2) integer in the following way. M randomly selects an n-bit integer r_i. If $r_i \notin Z_x^{+1}$ then he sets $\rho_i = r_i$ otherwise he randomly selects an integer $u_i \in Z_x^*$ and sets ρ_i equal to $u_i^2 \bmod x$ or $-u_i^2 \bmod x$ according to the outcome of a fair coin. In this way, any ρ_i belonging to Z_x^{+1} is a quadratic residue or a quadratic non residue with the same probability, and M can easily give a square root of ρ_i or $-\rho_i \bmod x$. In the second phase M generates the random integers $\sigma_1, \ldots, \sigma_{10n}$ and simulates the proof that each quadratic residue modulo x have a fourth root modulo x. M randomly selects an integer s_i. If $s_i \notin Z_x^{+1}$ then he sets $\sigma_i = s_i$, otherwise he randomly selects an integer $v_i \in Z_x^*$ and sets σ_i equal to $v_i^4 \bmod x$ or $-v_i^4 \bmod x$ according to the outcome of a fair coin. In this way, the σ_i's belonging to Z_x^{+1} are either quadratic residues or quadratic non residues with probability $1/2$; moreover M can easily give a fourth root of σ_i or $-\sigma_i \bmod x$.

Now we give a formal description of M.

Instructions for M.

Input: $x \in$ BL, where $|x| = n$.

1. Set Proof= empty string.
2. **Phase I:** (*Generation of $\rho_1, \ldots, \rho_{50n}$ and simulation of the proof that $-1 \bmod x$ is a quadratic non residue and x is a Regular(2) integer.*)
   For $i = 1$ to $50n$,
       randomly choose an n-bit integer r_i;
       if $r_i \notin Z_x^{+1}$ then set $\rho_i \leftarrow r_i$ and $u_i \leftarrow 0$;
       if $r_i \in Z_x^{+1}$ then
           randomly choose an integer $u_i \in Z_x^*$ and a bit b_i;
           set $\rho_i \leftarrow (-1)^{b_i} u_i^2 \bmod x$;
       append (ρ_i, u_i) to Proof.
3. **Phase II:** (*Generation of $\sigma_1, \ldots, \sigma_{10n}$ and simulation of the proof that each quadratic residue modulo x has a fourth root modulo x.*)
   For $i = 1$ to $10n$,
       randomly choose an n-bit integer s_i;
       if $s_i \notin Z_x^{+1}$ then set $\sigma_i \leftarrow s_i$ and $v_i \leftarrow 0$;
       if $s_i \in Z_x^{+1}$ then
           randomly choose an integer $v_i \in Z_x^*$ and a bit c_i;
           if HEAD set $\sigma_i \leftarrow (-1)^{c_i} v_i^4 \bmod x$;
       append (σ_i, v_i) to Proof.
4. Set $\tau = \rho_1 \circ \cdots \circ \rho_{50n} \circ \sigma_1 \circ \cdots \circ \sigma_{10n}$ and **Output:** (τ, Proof).

Lemma 6. The simulator M runs in probabilistic polynomial time and the probability spaces $M(x)$ and $View_B(x)$ are equal for each $x \in$ BL.

Proof. It is easy to see that the simulator runs in probabilistic polynomial time. Now we just have to prove that the random variable given by the output of M and the random variable representing the view of B in the protocol have the same distribution. First, we see that τ is randomly distributed among the $60n^2$-bit long strings. Moreover, if $\rho_i \in Z_x^{+1}$, the corresponding u_i is a random square root of ρ_i with probability $1/2$ or a random square root of $-\rho_i \bmod x$ with probability $1/2$, and if $\sigma_i \in Z_x^{+1}$, the corresponding v_i is a random fourth root of σ_i with probability $1/2$ or a random fourth root of $-\sigma_i \bmod x$ with probability $1/2$. $\qquad\qquad\square$

We have thus proved the following:

Theorem 7. (A,B) is a Non-Interactive Perfect Zero-Knowledge Proof System for the language BL.

5   Non-Interactive Electronic Cash Systems

In this section we introduce the concept of non-interactive electronic cash system. Let $\{U_i\}$ be a collection of users and B be a distinguished player called the bank.

A non-interactive electronic cash system is a multi-party protocol between B and $\{U_i\}$. It consists of five subprotocols: a start-up protocol executed by B before all operations take place, a protocol between a user U and B in which U opens an account, a withdrawal protocol between U and B, a spending protocol between two users U_i and U_j, and a deposit protocol between U and B. Moreover, each of these protocols requires no interaction between the parties involved. In fact, in a non-interactive electronic cash systems all the operations allowed to any party are the following: writing on a public file, sending a message to another party, and accepting of rejecting a message received by another party.

As real-life coins are physical objects, they guarantee a satisfactory level of security to a bank and to the users. On the other hand, an electronic cash system should meet the following two requirements: (a) support transactions and operations representing the digital equivalent of (at least) all the real-life operations, and (b) provide (at least) the same level of security guaranteed by its physical counterpart.

As requirement (a) is concerned, a non-interactive electronic cash systems supports the following operations: opening an account, withdrawing and depositing a coin, spending a coin or any desired fraction of a coin. We require that all these operations be performed without any interaction between the parties.

For requirement (b), a non-interactive electronic cash system addresses the following security requirements:

1. UNFORGEABILITY:
   a user should not be able to forge a spendable coin if the Bank does not withdraw the coin from his account;
2. SINGLE SPENDING:
   a user should not be able to spend twice a coin, even to different users;
3. NO FRAMING:
   if a coin has been spent only once, no coalition between the bank and other users should allow the bank to frame a user as a double spender of that coin;
4. UNTRACEABILITY:
   when a party receives a coin, he should not been able to trace it back to the users involved in spending protocols of that coin.

6   Our System

In this section we present our non-interactive electronic cash system. We describe the protocols for the following operations: start-up of the system, withdrawing a coin, spending a coin, spending a fraction of a coin, depositing a coin. The system requires the presence of a public file. It will be used by the parties to store some reference keys and random strings that will be useful to prove the correctness of their operations.

6.1   Start-up of the cash system

Let m be the number of different values that a coin can have: we assume that the value of a single coin is a power of 2, that is $v_t = 2^{t-1}$ for some $t \in \{1, \ldots, m\}$. Let

n be a security parameter and ID_B be an n^2-bit string denoting the identity of the Bank and σ, τ, ρ three sufficiently long random strings. Moreover, by (A,B) we denote the non-interactive perfect zero-knowledge proof system of membership to the language BL described in the previous section. Let c be a constant such that the random reference string used in the protocol (A,B) is n^c-bits long. At the start-up of the system and before opening any account, the Bank executes the following "start-up" protocol:

Start-up

- Let $\sigma = \sigma_1 \circ \cdots \circ \sigma_m$, where each σ_i is n^c-bits long.
- For $i = 1, \ldots, m$,
  randomly choose two n-bit prime numbers p_i, q_i such that $p_i \equiv q_i \equiv 3 \bmod 4$;
  compute $x_i = p_i q_i$;
  run the algorithm A on input x_i, p_i, q_i and σ_i obtaining as output Pf_i.
- Write $(x_1, \ldots, x_m, Pf_1, \ldots, Pf_m, ID_B, \sigma, \tau, \rho)$ on the public file.

6.2 Opening the account

This protocol is executed only once for each user, when he wants to open an account. First of all a user U verifies the proofs written by the Bank on the public file in the start-up protocol. Let ID_u be an n^2-bits string denoting the identity of U; then he writes on the public file ID_u, a Blum integer x_u of which he only knows the factorization and a (non-interactive) proof that x_u is a Blum integer. The Bank verifies this proof and checks that the identity ID_u is different from those already present in the public file.

Opening the account

U's computation:

- For $i = 1, \ldots, m$,
  run the algorithm B on input x_i and σ_i; if convinced, continue.
- Randomly choose two n-bit prime numbers p_u, q_u such that $p_u \equiv q_u \equiv 3 \bmod 4$ and compute $x_u = p_u q_u$.
- Run the algorithm A on input x_u and τ thus obtaining as output $Proof_u$.
- Write $(ID_u, x_u, Proof_u)$ on the public file.

Bank's computation:

- Run the algorithm B on input x_u and τ; if convinced, continue; else reject user U.
- Verify that ID_u is different from all the others identities.
- If all the verifications are successful, then accept user U else reject it.

6.3 Withdrawing a coin

Each user that has an account can ask for a coin of a certain value v_t to the Bank. In order to get a coin c of value v_t, a user U writes on the public file the Blum integer x_u and n randomly chosen quadratic residues $g_1, \ldots, g_n$ modulo x_t, where $1 \leq t \leq m$. Finally, he sends to the Bank the square roots of these quadratic residues in order to prove that he is the only user that is actually requesting the coin c. The Bank verifies that the square roots modulo x_t are correctly computed. The coin c will be the $(n+2)$-tuple $(g_1, \ldots, g_n, t; x_u)$.

Withdrawing a coin

U's computation:

- For $i = 1, \ldots, n$,
  randomly choose $h_i \in Z_{x_t}^*$ and compute $g_i = h_i^2 \bmod x_t$.
- Write the coin $c = (g_1, \ldots, g_n, t; x_u)$ on the public file and send $(h_1, \ldots, h_n)$ to the Bank.

Bank's computation:

- For $i = 1, \ldots, n$,
  verify that $g_i = h_i^2 \bmod x_t$.
- If all the verifications are successful, then accept the withdrawal of the coin else reject it.

6.4 Spending a coin

Given a Blum integer x and an n-tuple of integers $\vec{v} = (v_1, \ldots, v_n)$, we define the x-signature of $\vec{v}$ as the n-tuple of integers $\vec{r} = (r_1, \ldots, r_n)$ such that for $i = 1, \ldots, n$, it holds that if $v_i \in QR_x$ then $r_i^2 = v_i \bmod x$, else $r_i = 0$.

If x's factorization is given, then an x-signature of an n-tuple of integers can be efficiently computed in the following way: for $i = 1, \ldots, n$, if $v_i \in QR_x$ then randomly choose $r_i \in Z_x^*$ such that $r_i^2 = v_i \bmod x$; else set $r_i = 0$. Moreover, one can efficiently verify that an n-tuple of integers $\vec{r}$ is an x-signature of an n-tuple $\vec{v}$ of integers: for $i = 1, \ldots, n$, if $v_i \in Z_x^{+1}$ then verify that $r_i^2 = v_i \bmod x$.

We first describe the spending protocol in the case in which U_i spends the whole coin that has withdrawn from the Bank, and then consider the case in which he spends only a fraction of it.

Let $ID_{u_i} = d_1 \circ \cdots \circ d_n$, where $|d_l| = n$, and $\rho = \rho_1 \circ \cdots \circ \rho_n$, where $|\rho_i| = n$. In order to give a coin of value v_t to a user U_j, U_i sends the square roots modulo x_t of the published integers $g_1, \ldots, g_n$. Then he sends some information that will be useful to avoid the double-spending of the coin: the x_{u_i}-signature of the integers $w_1, \ldots, w_n$, where $w_l = \rho_l d_l g_l \bmod x_{u_i}$, for $l = 1, \ldots, n$. The signature is useful when a double-spending occurs: in fact it can be seen as a witness of the fact that U_i is spending the coin $(g_1, \ldots, g_n, t; x_{u_i})$ to user U_j.

Spending a coin

U_i's computation:

- For $l = 1, \ldots, n$,
  compute $w_l = \rho_l d_l g_l \bmod x_{u_i}$.
- Compute $\vec{s} = (s_1, \ldots, s_n)$, a x_{u_i}-signature of $\vec{w} = (w_1, \ldots, w_n)$.
- Send $(g_1, \ldots, g_n, t; x_{u_i}), (s_1, \ldots, s_n), (w_1, \ldots, w_n)$ to U_j.

U_j's computation:

- Verify that the coin $c = (g_1, \ldots, g_n, t; x_{u_i})$ has been written on the public file.
- For $l = 1, \ldots, n$,
  verify that $w_l = \rho_l d_l g_l \bmod x_{u_i}$.
- Verify that $\vec{s} = (s_1, \ldots, s_n)$ is a correct x_{u_i}-signature of $\vec{w} = (w_1, \ldots, w_n)$.
- If all the verifications are successful, then accept the coin else reject it.

6.5 Spending a fraction of a coin

Our goal is to allow user U_i to pass an arbitrary fraction of a coin to a certain user U_j and to allow user U_j to do the same with the coins received. In our scheme we require that each owner U_i of a coin $c = (g_1, \ldots, g_n, t; x_{u_i})$ of value $v_t = 2^{t-1}$, where $1 \leq t \leq m$, can compute two coins c_1 and c_2 of value $v_{t-1} = 2^{t-2}$, and so on recursively. Thus, to the coin c one can associate a complete binary tree $D_{c,t}$ of height t in which the root is c, the nodes at level 2 are two coins of value 2^{t-2}, and at level k there are 2^{k-1} coins each of value 2^{t-k+1}. In order to reach our goal, the tree $D_{c,t}$ has to satisfy the following properties: (a) before the execution of the spending protocol, U_i is able to obtain any coin he desires in the tree rooted with c, but any other U_j is not; (b) after U_i has given a coin d at level k to U_j, U_j is able to obtain any coin he desires in the tree rooted with d. Property (a) guarantees that a user U_i can spend a fraction of its coin to U_j, and property (b) allows to U_j to do the same with the coin received.

We propose the following scheme. In the public file a user U_i writes the coin $c = (g_1, \ldots, g_n, t; x_u)$ and also $m - 1$ pairs $(a_1, b_1), \ldots, (a_{m-1}, b_{m-1})$ of quadratic residues modulo x_t of which he only knows the square roots modulo x_t. Thus the tree $D_{c,t}$ can be computed by U_i in the following way: given a quadratic residue g associated to the root, U_i computes $ga_1 \bmod x_t$ and $gb_1 \bmod x_t$, that will be the quadratic residues associated to the two sons in the tree. Analogously, given a quadratic residue g associated to a node at level k, U_i computes the quadratic residues associated to the two sons at level $k + 1$ as $ga_k \bmod x_t$ and $gb_k \bmod x_t$. Now, suppose that the user U_i has received in some way (from the Bank or another user) a coin c of value v_t and wants to spend a coin of value v_k. First of all he computes the tree $D_{c,t}$ associated to c. Then, let d be a node at level $t - k + 1$ never considered before by U_i and let $g_1, \ldots, g_n$ be the quadratic residues modulo x_{t-k+1} associated to it. Then, to spend the coin d, U_i sends the square roots modulo x_t of $g_1, \ldots, g_n$, the square roots modulo x_t of the quadratic residues $a_{t-k+1}, \ldots, a_{m-1}$ and $b_{t-k+1}, \ldots, b_{m-1}$, and the x_{u_i}-signature computed as before.

6.6 Depositing a coin

The protocol for depositing a coin is exactly the same as that for spending a coin. That is, a user U will deposit a coin by spending it to the Bank. The Bank credits the coin on U's account if and only if she verifies that U has correctly spent the coin and that the same coin has not been already deposited by some other user. In this last case the Bank realizes that a double spending has occurred and has to determine the user that has spent the coin twice.

We give an informal description of a procedure for doing this. If a double-spending has occurred, then a user U has given the same coin to a user C_1 and another user D_1. Now, suppose that C_1 has given this coin to C_2, and so on until some C_l has given it to the Bank. Analogously, suppose that D_1 has given the same coin to D_2, and so on until some D_h has given it to the Bank (we assume for simplicity that $h = l$, as the case $h \neq l$ can be treated similarly). First of all observe that the Bank receives a signature $\bar{s}_1$ of $\bar{w}_1 = (w_{1,1}, \ldots, w_{1,n})$ and a signature $\bar{s}_2$ of $\bar{w}_2 = (w_{2,1}, \ldots, w_{2,n})$ that are computed from a same coin $(g_1, \ldots, g_n, t; x_u)$ but using two different identities $ID_1 = d_{1,1} \circ \cdots \circ d_{1,n}$ and $ID_2 = d_{2,1} \circ \cdots \circ d_{2,n}$. Notice that if this happens then the Bank realizes that a same coin has been given to two different users by seeing that the ratio $w_{1,i}/w_{2,i}$ is equal to $d_{1,i}/d_{2,i}$, for $i = 1, \ldots, n$. Now the Bank has to determine the author of the double spending. That is, she has to reconstruct the two paths $D_l, D_{l-1}, \ldots, D_1, U$ and $C_l, C_{l-1}, \ldots, C_1, U$ that have been taken (in the opposite direction) by the two copies of the coin during the spending protocols. For $i = l, l-1, \ldots, 1$, the Bank (a) proves to users D_i and C_i that a double spending has occurred (this can be done by giving to D_i and C_i all the signatures received in the spending protocols by users $D_{i+1}, \ldots, D_l$ and $C_{i+1}, \ldots, C_l$, and by the Bank), and (b) asks to users D_i and C_i who are the users D_{i-1} and C_{i-1} from whom they have received the coin (D_i and C_i can answer by simply giving the signatures received in the spending protocol).

This protocol stops when the Bank receives two signatures that have been sent by the same user U in two different spending protocols. In such a case, the Bank realizes that U is the author of the double spending protocol and can prove it using these two signatures. The protocol can stop before, if some user does not answer properly to the Bank; but in this case the Bank can consider that user as the author of the double spender. Moreover, the protocol works if a x_u-signature can be computed only by the owner of x_u's factorization. On the other hand, if x_u's factorization is not given, then an x_u-signature of an n-tuple of integers $\bar{v}$ can be efficiently computed only if $v_l \notin QR_{x_u}$ for $l = 1, \ldots, n$. However, as in the above procedure the v_l's are random and x_u is a Blum integer, the probability of this event is negligible (notice that if x_u has many prime factors, a randomly chosen integer in $Z_{x_u}^*$ has a negligible probability of being a quadratic residue modulo x_u, and thus it is possible to efficiently give a correct x_u-signature even if x_u's factorization is not given).

Remark. It is possible to avoid all the interaction needed in this protocol by changing the spending protocol in the following way: after that a user U_j has received a coin, he immediately deposits it to the Bank. In such a way, the Bank can detect immediately a double spending by simply checking if she has already received a signature computed on the same coin (but using different identities) of which U_j has sent an x_{u_j}-signature.

7 Proof of correctness

In this section we briefly give a proof of the four properties of our electronic cash systems: unforgeability, single spending, no framing and untraceability. We have the following:

Property 1. (UNFORGEABILITY) In our electronic cash system any user is not able to forge a spendable coin if the Bank does not withdraw the coin from his account.

Proof's sketch: Suppose U_i does not own a coin of a certain value; if he tries to spend it using the identity and the modulus of another user U_k, then he has to compute a correct signature modulo x_{u_k}, but this means that he is able to extract square roots modulo x_{u_k}. Moreover, if a user tries to withdraw a coin of a certain value v_t, he has to publish some quadratic residues and a value t; thus, if he does not own a coin of such a value, the user is easily caught by the Bank. □

Property 2. (SINGLE SPENDING) In our electronic cash system any user is not able to spend twice a coin, even to different users.

Proof's sketch: If a same coin is spent twice, then the Bank can realize that a double spending has occurred as she receives two signatures of a same coin. Moreover, the Bank identifies the author of the double-spending by running the procedure described in the previous section. □

Property 3. (NO FRAMING) In our electronic cash system if a coin has been spent only once, no coalition between the bank and other users should allow the bank to frame a user as a double spender of that coin.

Proof's sketch: If a user U spends a coin only once, then the Bank (even if in cooperation with other users) cannot frame U as a multiple spender. In fact, suppose the Bank wants to frame U as a multiple spender, then she has to exhibit two x_u-signatures on a same coin but usung two different n-tuple of integers (referring to the two different identities), but this means that he is able to extract square roots modulo x_u. □

Property 4. (UNTRACEABILITY) In our electronic cash system when a party receives a coin, he should not been able to trace it back to the users involved in spending protocols of that coin.

Proof's sketch: This can be easily seen from the fact that in a spending protocol a user sends only messages already written on a public file or messages that are depending only on itself and on the receiver. □

We refer the reader to the final version for formal proofs.

Acknowledgements

Many thanks go to Alfredo De Santis and Giuseppe Persiano for useful discussions and encouragements.

References

1. J. Bar Ilan and D. Beaver, *Non-Cryptographic Fault-Tolerant Computation in a Constant Number of Rounds of Interaction*, in Proc. of the 8th PODC (1989) pp. 201–209.
2. D. Beaver, S. Micali, and P. Rogaway, *The Round Complexity of Secure Protocols*, Proceedings of the 22nd Annual Symposium on the Theory of Computing, 1990, pp. 503–513.
3. M. Bellare and M. Yung, *Certifying Cryptographic Tools: The case of Trapdoor Permutations*, in Proceedings of CRYPTO-92.
4. M. Ben-Or, O. Goldreich, S. Goldwasser, J. Hastad, S. Micali, and P. Rogaway, *Everything Provable is Provable in Zero Knowledge*, in "Advances in Cryptology - CRYPTO 88", Ed. S. Goldwasser, vol. 403 of "Lecture Notes in Computer Science", Springer-Verlag, pp. 37–56.
5. J. Boyar, K. Friedl, and C. Lund, *Practical Zero-Knowledge Proofs: Giving Hints and Using Deficiencies*, Journal of Cryptology, n. 4, pp. 185–206, 1991.
6. M. Blum, A. De Santis, S. Micali, and G. Persiano, *Non-Interactive Zero-Knowledge*, SIAM Journal of Computing, vol. 20, no. 6, Dec 1991, pp. 1084–1118.
7. M. Blum, P. Feldman, and S. Micali, *Non-Interactive Zero-Knowledge and Applications*, Proceedings of the 20th Annual ACM Symposium on Theory of Computing, 1988, pp. 103–112.
8. R. Boppana, J. Hastad, and S. Zachos, *Does co-NP has Short Interactive Proofs ?*, Information Processing Letters, vol. 25, May 1987, pp. 127–132.
9. D. Chaum, *Security without Identification: Transaction System to Make Big Brother Obsolete*, in Communication of the ACM, 28, 10, 1985, pp. 1030–1044.
10. D. Chaum, A. Fiat, and M. Naor, *Untraceable Electronic Cash*, in "Advances in Cryptology - CRYPTO 88", Ed. S. Goldwasser, vol. 403 of "Lecture Notes in Computer Science", Springer-Verlag, pp. 319–327.
11. I. B. Damgard, *Payment Systems and Credential Mechanisms with Provable Security Against Abuse by Individuals*, in "Advances in Cryptology - CRYPTO 88", Ed. S. Goldwasser, vol. 403 of "Lecture Notes in Computer Science", Springer-Verlag, pp. 328–335.
12. A. De Santis, G. Di Crescenzo, and G. Persiano, *The Knowledge Complexity of Quadratic Residuosity Languages*, to appear in Theoretical Computer Science.
13. A. De Santis, G. Di Crescenzo, and G. Persiano, *Secret Sharing and Perfect Zero-Knowledge*, to appear in Proceedings of CRYPTO 93.

14. A. De Santis, S. Micali, and G. Persiano, *Non-Interactive Zero-Knowledge Proof-Systems*, in "Advances in Cryptology – CRYPTO 87", vol. 293 of "Lecture Notes in Computer Science", Springer Verlag.

15. A. De Santis, and G. Persiano, *Communication Efficient Zero-Knowledge Proof of knowledge (with Application to Electronic Cash)*, in Proceedings of STACS 92, pp. 449–460.

16. U. Feige, D. Lapidot, and A. Shamir, *Multiple Non-Interactive Zero-Knowledge Proofs Based on a Single Random String*, in Proceedings of 22nd Annual Symposium on the Theory of Computing, 1990, pp. 308–317.

17. N. Ferguson, *Single Term Off-Line Coins*, to appear in Proceedings of Eurocrypt 93.

18. L. Fortnow, *The Complexity of Perfect Zero-Knowledge*, in Proceedings of the 19th Annual ACM Symposium on Theory of Computing, 1987, pp. 204–209.

19. M. Franklin and M. Yung, *Secure and Efficient Off-Line Digital Money*, in Proceedings of ICALP 93.

20. O. Goldreich and E. Kushilevitz, *A Perfect Zero Knowledge Proof for a Decision Problem Equivalent to Discrete Logarithm*, in "Advances in Cryptology - CRYPTO 88", Ed. S. Goldwasser, vol. 403 of "Lecture Notes in Computer Science", Springer-Verlag.

21. O. Goldreich, S. Micali, and A. Wigderson, *Proofs that Yield Nothing but their Validity and a Methodology of Cryptographic Design*, Proceedings of 27th Annual Symposium on Foundations of Computer Science, 1986, pp. 174–187.

22. S. Goldwasser, S. Micali, and C. Rackoff, *The Knowledge Complexity of Interactive Proof-Systems*, SIAM Journal on Computing, vol. 18, n. 1, February 1989.

23. R. Impagliazzo and M. Yung, *Direct Minimum Knowledge Computations* "Advances in Cryptology – CRYPTO 87", vol. 293 of "Lecture Notes in Computer Science", Springer Verlag pp. 40–51.

24. I. Niven and H. S. Zuckerman, *An Introduction to the Theory of Numbers,* John Wiley and Sons, 1960, New York.

25. T. Okamoto and K. Ohta, *Disposable Zero-knowledge authentications and their Applications to Untraceable Electronic Cash*, in "Advances in Cryptology - CRYPTO 89", vol. 435 of "Lecture Notes in Computer Science", Springer-Verlag, pp. 481–496.

26. M. Tompa and H. Woll, *Random Self-Reducibility and Zero-Knowledge Interactive Proofs of Possession of Information*, Proc. 28th Symposium on Foundations of Computer Science, 1987, pp. 472–482.

A Unified Scheme for Routing
in Expander Based Networks

Shimon Even and Ami Litman

Computer Science Department
Technion, Israel Institute of Technology

Abstract. We propose a scheme for constructing expander based networks, by starting with a skeleton (directed and weighted) graph, and fleshing out its edges into expanders. Particular cases which can be built this way are: the multibutterfly, the multi-Beneš, a certain fat-tree and a superconcentrator.

The problem of on-line deterministic routing through any expander based network is shown to be solvable by a generalization of Upfal's algorithm.

1 Introduction

Several interconnection networks, whose structure is based on expanders, have been invented, and the problems of routing packets through them have been investigated. Our observation is that it may not be necessary to investigate each of these networks separately. We propose a scheme for constructing networks by starting with a skeleton (directed and weighted) graph, and fleshing out its edges into expanders. Particular cases which can be built this way are:

- The multibutterfly of Upfal [5],
- The multi-Beneš of Leighton et al. [2, 1],
- A fat-tree of Leiserson [3], and
- The superconcentrators of Pippenger [4].

Furthermore, we show that the problem of on-line deterministic packet routing through such an expander based network can be treated in the general case. That is, under certain conditions (on the load), a generalization of Upfal's algorithm can route N packets in time $O(\log N + L)$, where L bounds the distance packets have to travel.

So far, we have been unable to handle pipelining batches of packets, in the general case.

2 Expander Based Networks

A bipartite graph $\langle A, B, E \rangle$ is an (α, β) *expander* if every subset of A of cardinality $k \leq \alpha|A|$ is adjacent to at least $k\beta$ members of B. Let the ratio $\gamma = |A|/|B|$ be called the *concentration factor* of the expander. Some authors refer to expanders with $\gamma > 1$ as *partial concentrators*, but we will not make such a distinction. Also, we allow β to be less than 1.

It is known [5] that for any α, β and γ satisfying $\alpha\beta\gamma < 1$ there is a d such that for any positive integers a and b with ratio $a/b \leq \gamma$ there is an (α, β) expander, $\langle A, B, E \rangle$, such that the degree of its vertices is bounded by d, $|A| = a$ and $|B| = b$.

A *skeleton* is a scheme describing how to construct an expander-based interconnection network. Formally, it is a directed weighted graph, $G = \langle V, E, C \rangle$, where $C : V \Rightarrow \{\text{positive integers}\}$. The elements of G are called *skeleton vertices* and *skeleton edges*, and $C(v)$ is the *capacity* of the vertex v. The skeleton may be cyclic, but it has neither parallel edges nor self-loops. The skeleton is fleshed out to yield an actual directed graph $\bar{G} = \langle \bar{V}, \bar{E} \rangle$ (with *actual vertices* and *actual edges*) as follows. For each skeleton vertex v there corresponds a *pile* $\bar{v} \subset \bar{V}$ of $C(v)$ actual vertices. All these piles are disjoint, and their union is $\bar{V}$. For each skeleton edge $v \to u$, the actual graph contains an expander directed from $\bar{v}$ to $\bar{u}$. The union of all these expanders is $\bar{E}$. The α and β parameters of these expanders may vary. To be specific, each edge of E should be marked with its own α and β; however, these parameters will be determined later by other means.

Most of the known expander-based interconnection networks can be constructed as actual graphs of simple skeletons. Figure 1 depicts the skeleton of Upfal's multibutterfly [5]. Figure 2 shows the skeleton of a multi-Beneš network used by Leighton et al. [1, 2].

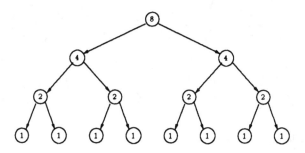

Fig. 1. The skeleton of an 8-input multibutterfly.

Figure 3 describes the skeleton of a fat-tree of Leiserson [3]. The original description of this fat-tree [3] used the terms *channel, capacity* and *wire*. In our terminology, a channel is a skeleton vertex, the capacity of the channel is the capacity $(C(v))$ of this vertex and a wire is an actual vertex. These figures reveal that a multibutterfly is a subgraph of a fat-tree and a wrapped multi-Beneš network is "almost" a subgraph of a fat-tree.

Figure 4 shows the skeleton of Pippenger's linear-sized superconcentrator [4]. The integer sequence there satisfies $n_{i+1} = 4\lceil n_i/6 \rceil$. All the diagonal edges rep-

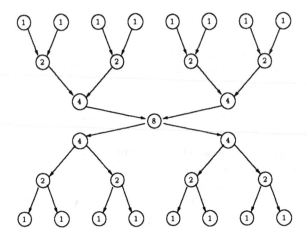

Fig. 2. The skeleton of an 8-input multi-Beneš network.

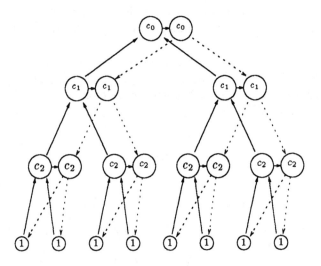

Fig. 3. The skeleton of an 8-leaves fat-tree.

resents $(\frac{1}{2}, 1)$ expanders. All the vertical edges, but the rightmost one, represent perfect matchings. We can consider these to be $(1, 1)$ expanders. The rightmost vertical edge, which always connect two vertices of capacity 8, represents a complete bipartite graph. We can consider this graph to be a $(\frac{1}{8}, 8)$ expander.

Henceforth, we use another approach to construct the actual edges as follows. Pick α and β satisfying $\alpha\beta < 1$. For a skeleton vertex u, let $\chi(u)$ denote the set of actual vertices preceding $\bar{u}$. Let $\gamma = |\chi(u)|/|\bar{u}|$. Connect $\chi(u)$ to $\bar{u}$ via an $(\alpha/\gamma, \beta)$ expander.

Note that by this construction, every subset of $\chi(u)$ of cardinality $k \leq \alpha|\bar{u}|$ is adjacent to at least $k\beta$ members of $\bar{u}$. Also note that for each skeleton edge

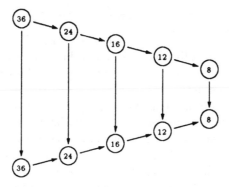

Fig. 4. The skeleton of Pippenger's linear-sized superconcentrator with 36 inputs.

$v \to u$, the actual edges from $\bar{v}$ to $\bar{u}$ form an (α', β') expander with $\beta' = \beta$ and $\alpha' \geq \alpha/\gamma$.

Unfortunately, the expanders constructed so far are useless for small sets. If $C(u) < 1/\alpha$ then the expander connecting $\chi(u)$ to $\bar{u}$ may be empty. To assure that each vertex of $\chi(u)$ has at least one edge leading to a vertex of $\bar{u}$, we make the following modification. When $C(u) < 1/\alpha$ we connect $\chi(u)$ to $\bar{u}$ by a $(\frac{1}{|\chi(u)|}, 1)$ expander; i.e., any bipartite graph where the out degree of every vertex of $\chi(u)$ is at least one. We refer to these graphs as *special* expanders.

3   Routing

We consider a variant of the on-line deterministic packet routing problem on an actual graph $\bar{G}$ generated from a skeleton G. In this variant, each packet carries a directed path of G it is supposed to trace. We refer to this path as the *trail* of the packet. Note that the trail is defined in terms of the skeleton, but the motion occurs in the expanded graph; the actual vertex the packet passes through, in a pile, is not specified. The algorithm is allowed to swap two packets residing on the ends of an edge. In that case, the corresponding skeleton edge is prefixed to the trail of the retracting packet.

In this paper we consider only the case where each vertex can store at most one packet. Our results, however, can be extended to the case where each vertex can store a constant number of packets.

We will use the concepts of *load* and *load factor* introduced by Leiserson [3]. Our definitions are more awkward, due to the issue of special expanders. Let M be a routing task. For a skeleton vertex u, define load(M, u) as the total number of times packets have to pass through $\bar{u}$. (The initial and terminal positions count. If a packet needs to pass through $\bar{u}$ several times, all passes are counted.) Define the adjustment function

$$\text{adj}(x) \triangleq \begin{cases} 0 & x = 1 \\ x & \text{otherwise} \end{cases}$$

The *load factor* of u is defined as $\lambda(M,u) \stackrel{\triangle}{=} \frac{\mathrm{adj}(\mathrm{load}(M,u))}{C(u)}$, and the load factor of the routing task M is $\lambda(M) \stackrel{\triangle}{=} \max\{\lambda(M,u) : u \in V\}$.

The rational for the adjustment is that a load of exactly one packet may seem significant for small piles, but is actually trivial. This single packet will be blocked by no other packet.

In our routing algorithm we use the concept of a *potential* of a packet: Let ω be a real number, whose choice will be discussed later, such that $0 < \omega < 1$. For a packet x, its potential at a certain moment is denoted by $P(x)$, which is either some integral power of ω, or 0. (Initially, it is positive.) When a packet moves a step forward, but has not reached its destination yet, its potential is multiplied by ω. If a packet has reached its destination, it *evaporates* and its potential is 0. If a packet moves backward, its potential is divided by ω.

We shall present a specific algorithm shortly. However, our results hold for all algorithms that have the following *Q-property*.

- A packet x retracts only if it is swapped with another packet, y, and this occurs only if $\omega P(y) > P(x)$ (before the swap).
- The algorithm works in *steps*, where each step takes $O(1)$ time.
- If a packet x residing in a vertex u does not move during a step, and there is an (actual) edge from u to v such that the corresponding skeleton edge is next on the trail of x, then there is a packet y that visits v during this step, and its potential (while in v) satisfies $P(y) \geq \omega P(x)$.

3.1   A Q-algorithm

The following algorithm is a generalization of Upfal algorithm [5], and has the Q-property. The algorithm uses a coloration of the actual edges. We assume that there is a fixed upper bound on the degree of the skeleton vertices, where both incoming and outgoing edges are counted. Since the degree of the vertices in each expander is bounded (by d), it follows that the degree of the actual vertices is bounded by a constant. Therefore, there is a constant, c, such that the actual edges can be colored, using at most c colors, and no two actual edges which share an end-point have the same color.

The algorithm is divided into *steps*. Each step is divided to c *substeps*. The operations performed in each step are as follows:

```
begin
    for 1 ≤ i ≤ c do
    begin
        for every edge u → v of color i in Ḡ parallel do
        begin
            if u contains a packet x and
                the skeleton edge which corresponds to u → v
                is next on the subsumed trail of x
            then do
            begin
                if v contains a packet y and ωP(x) > P(y)
                then swap packets x and y
                if v contains no packet then move x to v
            end
        end
    end
end
```

3.2   Analysis of a Q-algorithm

We shall use the following lemma.

Lemma 1. *Let $H = \langle A, B, E \rangle$ be an (α, β) expander, where $\beta > 1$ is a positive integer. Assume $A' \subset A$, such that $|A'| \leq \alpha|A|$. There is an assignment*

$$f : A' \Rightarrow 2^B ,$$

such that for every $a \in A'$: $f(a) \subseteq \Gamma(a)$, $|f(a)| = \beta$, and if $a \neq b$ then $f(a) \cap f(b) = \emptyset$.

Proof. The proof follows from Hall's theorem, either by induction, or by first magnifying the set A' by a factor β, finding a matching of the magnified set into B, and shrinking A' back to its previous size.

□

Let us use the following convention. If $G : D \Rightarrow \{\text{reals}\}$, and $D' \subseteq D$, then

$$G(D') \stackrel{\triangle}{=} \sum_{x \in D'} G(x) .$$

Our aim is to analyze how one step of a Q-algorithm changes the potential of the whole system. Let X be the set of packets, and let $P_0(x)$ $(P_1(x))$ denote the potential of packet x before (after) the step is performed. The next lemma claims that a step reduces the system potential by, at least, a constant factor.

Lemma 2. *For every integral $\beta > 1$, there exist an $0 < \omega < 1$ and a $\sigma < 1$ such that, if a step of the algorithm is applied to a routing task M for which $\lambda(M) \le \alpha$, then*

$$P_1(X) \le \sigma P_0(X) .$$

Proof. Let us define $P_\Delta(X) \overset{\triangle}{=} P_0(X) - P_1(X)$, and prove that there exist a constant $\epsilon_1 > 0$ such that

$$P_\Delta(X) \ge \epsilon_1 P_0(X) . \tag{1}$$

Let us divide X to two subsets:

- S — the set of stationary packets; i.e. packets that have not moved during the step.
- M — the set of mobile packets; a mobile packet may end up in its start point, but it has moved during the step.

Let $u \in V$ be a skeleton vertex, and let $S' \subseteq S$ be the set of stationary packets residing in the vertices of $\chi(u)$ and needing to move into $\bar{u}$. By the load assumption, which remains valid even when the trails change according to the movements of the packets, $|S'| \le \alpha C(u)$. (Note that this inequality holds also when $\alpha C(u) < 1$; in this case the expander is a special one and $S' = \emptyset$.) Thus, by Lemma 1, each packet $x \in S'$ can be assigned its own subset of vertices $U_x \subset \bar{u}$, such that $|U_x| = \beta$ and these subsets are disjoint.

An actual vertex v is called *active* if, during the step, some packet entered it or left it. For an active vertex v, let $y(v)$ be a packet of maximum potential among those which visited v during the step.

We use a book-keeping argument of potential redistribution; i.e., we show that there is a way to redistribute $P_0(X)$ among $X \cup \bar{V}$ in such a way that:

The (virtual) potential of any packet is reduced by at least a constant fraction.

- The virtual potential given to any vertex v is less than the reduction of the real potential caused by packets advancing into or out of v during this step.

This potential redistribution has no effect on the algorithm; it is used only to analyze the algorithm. We express the potential redistribution by a function

$$F : X \times (X \cup \bar{V}) \Rightarrow R ,$$

where $F(x, y)$ denotes the virtual potential transferred from packet x to packet or vertex y.

$F(x, y)$ is defined to be 0, except in the following cases:

1. $x, y \in S$ and packet y is in a vertex of U_x: $F(x, y) \overset{\triangle}{=} \epsilon_2 P_0(x)$,
2. $x \in S$ and vertex $y \in U_x$ is active: $F(x, y) \overset{\triangle}{=} \epsilon_2 P_0(x)$,

3. $x \in M$ and packet x is in vertex y in the beginning of the step: $F(x,y) \overset{\triangle}{=} \epsilon_2\epsilon_3 P_0(x)$.

The value of the constants $\epsilon_2, \epsilon_3 > 0$ will be discussed later. Part (3) of F's definition implies that

$$F(M, \bar{V}) = \epsilon_2\epsilon_3 P_0(M) . \tag{2}$$

Parts (1) and (2) imply that for every $x \in S$

$$F(x, S \cup \bar{V}) = \beta\epsilon_2 P_0(x) .$$

Next, let us investigate the transfers to $x \in S$. If $F(y, x) > 0$ then, by part (1), both packets are stationary, and therefore, $\omega P_0(y) \leq P_0(x)$. Thus,

$$F(y, x) = \epsilon_2 P_0(y) \leq \epsilon_2 \frac{1}{\omega} P_0(x) .$$

Also, x resides in a vertex of U_y and no other packet transfers any virtual potential to x. Thus,

$$F(S, x) \leq \epsilon_2 \frac{1}{\omega} P_0(x) .$$

It follows that the net amount of virtual potential which is transferred from x is

$$F(x, S \cup \bar{V}) - F(S, x) \geq \beta\epsilon_2 P_0(x) - \epsilon_2 \frac{1}{\omega} P_0(x)$$

$$= (\beta - \frac{1}{\omega})\epsilon_2 P_0(x) .$$

Let us choose ω close to 1 and $\epsilon_3 > 0$ such that $\beta - \frac{1}{\omega} \geq \epsilon_3$. Therefore,

$$F(x, S \cup \bar{V}) - F(S, x) \geq \epsilon_2\epsilon_3 P_0(x) .$$

Hence, the virtual potential of any stationary packet is reduced by at least a fraction of $\epsilon_2\epsilon_3$. Summing over all elements of S, one gets

$$F(S, S \cup \bar{V}) - F(S, S) \geq \epsilon_2\epsilon_3 P_0(S) ,$$

which means that

$$F(S, \bar{V}) \geq \epsilon_2\epsilon_3 P_0(S) . \tag{3}$$

Summing up Equation 2 and Equation 3, one gets

$$F(X, \bar{V}) \geq \epsilon_2\epsilon_3 P_0(X) . \tag{4}$$

In order to achieve our target, Equation 1, we want to bound $F(X, \bar{V})$ from above, by a constant times $P_\Delta(X)$.

Let us call a movement of a packet or a packet swap an *event*. Each event causes a reduction in the potential. The total reduction of the potential of the system is the sum of the reductions of all the events in the step. If in an event e, packet y moves forward, and its potential before the event has been $P(y)$, then the reduction of the potential associated with e, $\Delta(e)$, satisfies

$$\Delta(e) \geq P(y) \cdot f(\omega) , \tag{5}$$

where $f(\omega) > 0$. (To be exact, $f(\omega) = (\omega - 1)^2$.)

For $u \in \bar{V}$ there are at most two packets x for which $F(x, u) \neq 0$; one, x_1, stationary (per part (2)) and one x_2, mobile (per part (3)). In any case, if there is such an x then u is active during the step. Let us look at the packet y which satisfies $y = y(u)$, and associate with u an event in which y moves to u or from it.

$$F(X, u) \leq F(x_1, u) + F(x_2, u)$$
$$= \epsilon_2 P_0(x_1) + \epsilon_2 \epsilon_3 P_0(x_2) \,. \tag{6}$$

Since x_1 is stationary, the packet z, which was in u and prevented the advance of x_1 to u, satisfies

$$\omega P_0(x_1) \leq P(z) \,,$$

and the choice of y implies that $P(z) \leq P(y)$. Thus, $\omega P_0(x_1) \leq P(y)$, and we have,

$$P_0(x_1) \leq \frac{1}{\omega} \cdot P(y) \,. \tag{7}$$

If there is an x_2, its move has started from u. Thus, $P_0(x_2) \leq P(y)$. Together with Equations 6 and 7, this yields

$$F(X, u) \leq (\epsilon_2 \cdot \frac{1}{\omega} + \epsilon_2 \epsilon_3) P(y) \,,$$

and by Equation 5,

$$F(X, u) \leq (\epsilon_2 \cdot \frac{1}{\omega} + \epsilon_2 \epsilon_3) \frac{1}{f(\omega)} \cdot \Delta(e) \,.$$

One gets,

$$F(X, u) \leq \epsilon_2 g(\epsilon_3, \omega) \cdot \Delta(e) \,,$$

where $g(\epsilon_3, \omega)$ denotes the proper positive function. Let us choose ϵ_2 satisfying: $\epsilon_2 g(\epsilon_3, \omega) \leq 1$. Thus,

$$F(X, u) \leq \Delta(e) \,,$$

and the virtual potential transferred to u is less than or equal to $\Delta(e)$. Since each event is associated with at most two different vertices, it follows that

$$F(X, \bar{V}) \leq 2 \cdot P_\Delta(X) \,.$$

Joining this with Equation 4, one gets

$$2P_\Delta(X) \geq \epsilon_2 \epsilon_3 P_0(X) \,,$$

and with the proper choice of $\epsilon_1 > 0$ it yields

$$P_\Delta(X) \geq \epsilon_1 P_0(X) \,.$$

$\square$

Theorem 3. *Assume $\beta > 1$ is an integer. Let M be a routing task of N packets satisfying $\lambda(M) \leq \alpha$, and let L bound the trails length. Then the routing algorithm will perform M in time $O(\log N + L)$.*

Proof. Trivial.

$\square$

4 Extensions

We proved lemma 2 and theorem 3 for integral $\beta > 1$. By using a lemma analogous to 1 for a rational β, we can extend lemma 2, and hence theorem 3 for any rational $\beta > 1$.

Under some conditions our results hold also for a routing task M with a load factor of $\alpha(\beta + 1)$. Assume that for any skeleton vertex u, $\alpha C(u)$ is either an integer or less than one, and furthermore that we use complete bipartite graphs for special expanders. In this case, our results hold for a routing task M with $\lambda(M) \leq \alpha(\beta + 1)$. (Note that in this case we may have $\lambda(M) > 1$.)

In our proofs, we used the load condition, $\lambda(M) \leq \alpha$, only in lemma 2 as follows. We defined S' as the set of stationary packets residing in $\chi(u)$, and concluded that $|S'| \leq \alpha C(u)$. However, we can reach the same conclusion from the weaker condition of $\lambda(M) \leq \alpha(\beta + 1)$ as follows.

Let us consider two cases. In the first case $\alpha C(u) < 1$ and the expander exiting $\chi(u)$ is a special one; i.e., a complete bipartite graph. Assume $S' \neq \emptyset$. Then all the vertices of $\bar{u}$ are "busy" during the step and it can be shown that this requires a load of at least $C(u)$ packets. Together with the stationary packets of S', we have

$$\lambda(M, u) \geq \frac{C(u) + 1}{C(u)} = 1 + \frac{1}{C(u)} > \alpha\beta + \frac{1}{C(u)} > \alpha\beta + \alpha$$

A contradiction.

In the second case $\alpha C(u)$ is an integer and the expander exiting $\chi(u)$ is a regular one. Let $V' \subset \chi(u)$ be the set of vertices holding S'. Assume $|V'| > \alpha C(u)$. Using a subset of V' with $\alpha C(u)$ elements, it follows that $|\Gamma(V')| \geq \alpha\beta C(u)$. All the vertices of $\Gamma(V')$ are "busy" during the step, and this requires a load of at least $|\Gamma(V')|$ packets. This load, together with the stationary packets of S', is greater than $\alpha(\beta + 1)C(u)$. A contradiction.

In the general case of cyclic networks, we do not know how to pipeline several *batches* of packets as done by Upfal [5] and Leighton and Maggs [2] in the case of the multibutterfly. However, if the skeleton is a directed tree, with edges leading away from the root, we can pipeline several batches. In fact, we can do so even when the load factor of each batch is $\alpha(\beta+1)$ as discussed above. On the multibutterfly, any such batch can be a complete input to output permutation.

References

1. Arora, S., T. Leighton, B. Maggs, "On-line algorithms for path selection in a non-blocking network," *Proc. of the Twenty Second Annual ACM Symp. on Theory of Computing*, pp. 149-158, 1990.
2. Leighton, F.T. and B.M. Maggs, "Fast algorithms for routing around faults in multibutterflies and randomly-wired splitter networks," *IEEE Trans. Comp.*, vol. 41, pp. 578-587, 1992.
3. Leiserson, C.E, "Fat-Trees: Universal networks for hardware-efficient supercomputing," *IEEE Trans. Comp.*, vol. C-34, pp. 892-901, 1985.

4. Pippenger, N.J., "Superconcentrators," *SIAM J. Comput.*, vol. 6, pp. 298-304, 1977.

5. Upfal, E. "An $O(\log N)$ deterministic packet routing scheme," *JACM*, vol. 39, pp. 55-70, 1992.

Dynamization of Backtrack-Free Search for the Constraint Satisfaction Problem*

Daniele Frigioni[2] Alberto Marchetti-Spaccamela[1] Umberto Nanni[2]

[1] Dipartimento di Informatica e Sistemistica, University of Rome "La Sapienza", via Salaria 113, I-00184 Rome, Italy
[2] Dipartimento di Matematica Pura ed Applicata, University of L'Aquila, via Vetoio loc. Coppito, I-67010 l'Aquila, Italy.

Abstract. Many AI tasks can be formulated as a *Constraint Satisfaction Problem* (CSP), i.e. the problem of finding an assignment of values for a set of variables subject to a given collection of *constraints*. In this framework each constraint is defined over a set of variables and specifies the set of allowed combinations of values as a collection of tuples.

In some cases the knowledge of the problem defined by the set of constraints may vary along the time. In particular one might be interested in further *restrictions* i.e. in deletions of values from existing constraints, or in introducing new ones.

In general the problem to find a solution to a CSP is NP-complete, but there exist some cases that can be solved efficiently. In this paper we consider classes of problems with a tractable solution, and present dynamic algorithms that solve this problem efficiently and are shown to be optimal.

1 Introduction

A *network of constraints* is defined by a set $V = \{v_1, v_2, \ldots, v_n\}$ of *variables*, each with a given domain D_i, and a set $E = \{e_1, e_2, \ldots, e_m\}$ of *constraints*, each defining the collection of allowed values for a given set of variables: $e_i(v_{i_1}, v_{i_2}, \ldots, v_{i_k}) \subseteq D_{i_1} \times D_{i_2} \times \ldots \times D_{i_k}$. The *Constraint Satisfaction Problem* (CSP) consists in determining a solution, i.e. an assignment of values $(\overline{v}_1, \overline{v}_2, \ldots, \overline{v}_n)$ such that all the constraints are satisfied, that is for any constraint e_i: $(\overline{v}_{i_1}, \overline{v}_{i_2}, \ldots, \overline{v}_{i_k}) \in e_i$.

This problem has been introduced by Montanari [11], and is easily shown to be NP-complete, but in particular cases, it is possible to solve it efficiently.

Also restricting the attention to binary CSP (where any constraint is defined over a *pair* of variables), the problem remains hard. It is straightforward to describe a binary CSP (V, E) in terms of a labeled *graph of constraints*, where the nodes coincides with the variables, and any constraint is represented by an undirected edge labeled by the set of allowed pairs. It is possible to express many classical problems, such as *Graph Coloring*, or *Scene Labeling* as binary CSP.

In the binary case, if the graph of constraints is indeed a tree, the problem becomes tractable, and polynomial solutions are provided by Freuder [7], and Detcher

* Work supported by the ESPRIT II Basic Research Action no.7141 (Alcom II) and by the Italian Project "Algoritmi e Strutture di Calcolo", Ministero dell'Università e della Ricerca Scientifica e Tecnologica.

and Pearl [5], who achieve an optimal time bound of $O(nk^2)$, corresponding to the size of the description of a binary CSP with n variables (and $m < n$ constraints), each defined over a domain of size at most k.

In some cases the knowledge of the problem defined by the set of constraints may vary along the time. For example many problems in *image processing* can be modeled as a CSP [11]. During the computation, as our knowledge of the problem increases, it may be the case that the universe of allowed solutions shrinks more and more. In other cases, it might be the structure of the constraints themselves that is changing upon the time. In this situations it is interesting to handle efficiently *restrictions*, i.e. either deletion of values from existing constraints, or introduction of new constraints, while maintaining information on the satisfiability of the constraints and a current solution.

We deal with dynamic binary CSP and present a dynamic solution requiring $O(nk^2)$ total time in a sequence of at most $O(nk^2)$ tuple deletions, thus achieving an optimal $O(1)$ amortized time bound. The case of more general restrictions (i.e. including definition of new constraints) is also considered. We follow an approach based on the *directional arc-consistency* property introduced by Detcher and Pearl [5], and in particular we implicitly maintain the set of all the admissible solution for a CSP. On request a solution (having size n) is provided in $O(n)$ worst case time.

Finally we address a class of (non binary) polynomial CSP, namely the *acyclic CSP's* showing that the approach of Detcher and Pearl and our dynamization are valid also for that larger class.

The rest of the paper is organized as follows. In section 2 we introduce a more formal definition of CSP and a description of properties and methods exploited to deal with this problem. In the following section 3 we present our approach to dynamic binary CSP.

2 Basic Definitions

The *Constraint Satisfaction Problem* (CSP) has been introduced by Montanari [11] to capture and study in a single formal framework a wide set of problems arising in various fields of computer science and combinatorics. CSP can be regarded in terms of *hypergraphs*, a well studied data structure (see, e.g., [3]).

Definition 1. A *labeled hypergraph* H_L is a triple $< N, A, l >$ where N is a finite set of *nodes*, A is a finite set of *hyperedges* such that $A \subseteq P^+(N)$ and $l : A \to L$ is a *labeling function* with values in the set of *labels* L.

In our context, it is convenient to partition the set of hyperedges A according their cardinality, so that $A_k = \{h | h \in A \text{ and } |h| = k\}$. It is possible to consider the restrictions of the labeling function l:

$$l_k : A_k \to L_k$$

with the resulting set of labels $L = \cup_{k=1}^n L_k$.

In the previous section we have given a first possible definition of a CSP as the problem of finding an assignment $\overline{V}$ of values for a set of variables V satisfying a given set of constraint E. A CSP or, more precisely, the *network of constraints* (V, E) which it is based on, can be modeled by means of a *constraint hypergraph*.

Definition 2. The *constraint hypergraph* of a *network of constraints* (V, E) is a labeled hypergraph $H_L =< N, A, l >$ where $N \equiv V$, $A = \{(v_{i_1}, v_{i_2}, \ldots, v_{i_k}) \mid e_i(v_{i_1}, v_{i_2}, \ldots, v_{i_k})$ is a constraint in $E\}$ and the labeling function l is such that any hyperedge in A is labeled with the corresponding constraint in E, and in particular: $l_k : A_k \rightarrow P(U^k)$ where U is a finite domain for all the variables in V.

A *solution* for a CSP is an assignment of values $\overline{V} = (\overline{v}_1, \overline{v}_2, \ldots, \overline{v}_n)$ to the n variables of the network that simultaneously satisfy all the constraints, i.e. such that for any hyperedge $h = (v_{i_1}, v_{i_2}, \ldots, v_{i_k}) \in A$: $(\overline{v}_{i_1}, \overline{v}_{i_2}, \ldots, \overline{v}_{i_k}) \in l(h)$.

In this paper we will be primarily concerned with binary CSP, i.e., CSP where each constraint involves at most two variables. Such a network will be called *constraint graph*. A binary constraint $e(v_i, v_j)$ is a subset of the cartesian product of their domains, i.e., $e(v_i, v_j) \subseteq D_i \times D_j$.

A binary constraint $e(v_i, v_j)$ is usually represented using a $k \times k$ matrix denoted as $R_{i,j}$, where $k = max\{|D_i|\}$, and the entries 0 and 1 indicate forbidden and permitted pairs of values, respectively. Thereafter we refer to a constraint as its binary matrix $R_{i,j}$.

A trivial (exponential time) algorithm to find a solution for a CSP is based on *backtracking*. In [7, 10, 11] there is shown that, if the CSP satisfies certain conditions on the structure of the constraint graph and on the values within the constraints, the problem turns out to be polynomially solvable.

Following their approach, which deals with directed graphs[3], we need to consider any binary constraint $R_{i,j}$ as labeling both the two directed arcs (v_i, v_j) and (v_j, v_i) whose matrix representations satisfy $R_{i,j} = R_{j,i}^T$.

Definition 3. A directed arc (v_i, v_j) is *consistent* iff for any value $x \in D_i$ there is a value $y \in D_j$ such that $(x, y) \in R_{i,j}$. A constraint graph is *arc-consistent* if each of its directed arcs is consistent.

In [6] Freuder defined the property of *k-consistency* as a generalization of arc-consistency. This property specifies that if we choose values for any $k - 1$ variables that satisfy the constraints on these variables, along with any k-th variable, it is possible to find a value for the k-th variable such that all the constraints on the k variables will be satisfied by the k values taken together. *Strong k-consistency* is k'-consistency for all $k' \leq k$.

An *ordered constraint graph* (G, d) is a constraint graph G in which the nodes are given a total order d. The ordering d corresponds to the order in which the variables are chosen for instantiation in the backtrack search.

Definition 4. The *width of a node* x in an ordered constraint graph (G, d) is the number of links (y, x) such that y precedes x in d. The *width of an ordering* is the maximum width of all nodes. The *width of a constraint graph* is the minimum width of all the orderings of the graph.

---

[3] In a directed graph $G =< V, E >$, any edge $(v_i, v_j) \in E \subseteq V \times V$ is an *ordered pair*.

Freuder in [7] also provided an $O(n^2)$ algorithm for finding both the width of a graph and the ordering corresponding to this width. He further showed that a constraint graph is a tree iff it has width 1. Backtracking occurs when an instantiation chosen during a backtrack search, consistent with all previous choices, must be discarded later in the search when no consistent instantiation can be made for a variable at a lower level in the backtrack tree. A search for the solution is called *backtrack-free* if it terminates without making backtracking steps.

The obvious interest in such an approach is due to the fact that a backtrack-free search can provide a solution in linear time. Freuder showed the following relation between width and consistency that guarantees backtrack-free search for the solution.

Theorem 5. [7] *There exists a backtrack-free search for a binary CSP if the level of strong consistency is greater than the width of the constraint graph.*

Detcher and Pearl noticed that full arc-consistency is more than what is actually required to achieve a backtrack-free solution for width-1 CSP [5]. The arc-consistency is required only w.r.t. a single direction, i.e. the one in which a backtrack search selects variables for instantiation. With this simple observation they motivate the following definition.

Definition 6. Given an ordered constraint graph (G, d), then G is *d-arc consistent* if all the edges directed along d are consistent.

They also proposed an algorithm for achieving directional arc-consistency for any ordered constraint graph (the ordering $d = (v_1, v_2, \ldots, v_n)$ is assumed). This algorithm applies a procedure called **Revise** to each arc (v_i, v_j) with $i \leq j$ following the given ordering d in the reverse order. Procedure $\textbf{Revise}(v_i, v_j)$, given in [10], deletes values from the domain D_i in $O(k^2)$ steps in the worst case ($k = \max\{|D_i|, |D_j|\}$), until the directed arc (v_i, v_j) is made consistent.

Detcher and Pearl proved also the following result.

Theorem 7. [5] *A tree-like CSP can be solved in $O(nk^2)$ steps, and this is optimal.*

As the graph coloring problem suggests, the general problem of finding a solution for a CSP, also in the binary case, is NP-complete. On the other hand the results of Mackworth [10] Freuder [7] and Detcher and Pearl [5] show that, if suitable properties hold (namely, if the network has a sufficient degree of consistency), the problem becomes tractable.

3 Dynamic binary CSP

In this section we describe a dynamic solution for acyclic binary CSP's while performing sequences of updates, consisting either in *restrictions* (deletion of values from existing constraints, or introduction of new constraints) or *relaxations* (insertion of values or deletions of constraints).

The idea behind this new approach is to maintain the property of d-arc consistency, which guarantees a backtrack-free search for the solution for tree-like problems, when performing separately arbitrary sequences of dynamic operations of the following kind:

1. pair deletion from preexisting constraints;
2. arc (constraint) insertion;
3. pair insertion in preexisting constraints;
4. arc (constraint) deletion.

We remark that deleting an arc (v_i, v_j) is equivalent to insert in $R_{i,j}$ all the pairs currently not contained in that constraint: in this way the constraint $R_{i,j}$ becomes what is called a *universal constraint*, i.e., $R_{i,j} = D_i \times D_j$. Analogous considerations regard operations 2 and 1.

Any operation of the kind 3 or 4 (inserting a pair in a preexisting constraint or deleting an arc) is a trivial operations to handle by itself, in the case of a previously satisfiable CSP. Though, if we add any pair in an unsatisfiable network (or if we perform an intermixed sequence of restrictions and relaxations), it is necessary to consider the newly introduced values: as it will be shown, this makes the fully dynamic problem harder.

3.1 Pair deletions

A trivial solution of the pair deletion problem consists in applying for each deletion the $O(k^2)$ off-line algorithm Revise to all the arcs involved in the modification, obtaining an $O(nk^2)$ time bound (worst-case) for each deletion[4].

In this section we propose a data structure and a dynamic algorithm that allow us to maintain the d-arc consistency for tree-like CSP in $O(nk^2)$ total time, under an arbitrary sequence of pair deletions, i.e., in $O(1)$ amortized time for each pair deletion.

The only case in which a pair deletion can modify the directional arc-consistency of an ordered tree-like network is when we delete a pair (a, b) from a constraint $R_{i,j}$ while the pair (a, b) is the only one that guarantees consistency of the arc (v_i, v_j) for the value $a \in D_i$. In other words, before the deletion of (a, b), for any $b' \in D_j$ such that $b' \neq b, R_{i,j}(a, b') = 0$.

Lemma 8. *Let* $(a, b) \in R_{i,j}$ *be a pair such that for any value* $b' \in D_j$, *with* $b' \neq b, R_{i,j}(a, b') = 0$. *Deleting such a pair* (a, b), *in order to restore arc consistency for edge* (v_i, v_j), *requires the deletion of the value* a *from the domain* D_i.

Proof. It is sufficient to observe that, after the deletion of (a, b), the value a is such that for any value $x \in D_j$, $R_{i,j}(a, x) = 0$, i.e., the arc (v_i, v_j) is not consistent, according to definition 3.

Let us call *critical* both a pair complying the hypotheses of lemma 8, and its deletion.

After the deletion of a critical pair $(a, b) \in R_{i,j}$, we have that restoring arc consistency implies the "deletion" of value a from domain D_i of variable v_i, which means that no solution for the considered CSP can assign $v_i = a$. This, in turn, implies that any other pair (z, a) in any other constraint $R_{h,i}$, for any z, h, can

---

[4] For simplicity we express bounds in terms of a parameter k, denoting the maximum size of any domain for the variables in V, i.e.: $k = \max_{v_i \in V}\{|D_i|\}$

be taken into consideration to build up solution for CSP or to support consistency of arcs. These considerations lead to the ideas behind our algorithms: to support explicit pair deletion, requested by the user, and *implicit* pair deletions, performed by the algorithms in order to restore consistency.

In the rest of this section we first describe our data structures, and then provide the details of algorithms.

The data structure. To simplify our description, we suppose to support only pair deletions from constraint. In this situation, as previously remarked, the domains of variables can only shrink. Let us denote as $D_i(0)$ $(i = 1, 2, \ldots, n)$ the *initial domain* for variable v_i. For any i, the following inequalities trivially hold: $|D_i| \leq |D_i(0)| \leq k$.

For each variable v_i we maintain the current content of the domain in a binary vector R_i, indexed by the element of the initial domain $D_i(0)$:

$$R_i(x) = \begin{cases} 1 \text{ if the value } x \text{ is still in } D_i \\ 0 \text{ if the value } x \text{ has been deleted from } D_i. \end{cases}$$

Moreover, for each variable v_i, we maintain an integer value d_i containing the current cardinality D_i.

The basic idea is to represent a binary constraint $R_{i,j}$ as an array of columns, any column being handled both as an array and as a set. In other words we maintain a collection of sets of values $\{x_1, x_2, \ldots, x_{k_w}\} \subseteq D_i$ each corresponding to a value $w \in D_j$, i.e., all the pairs $(*, w) \in R_{i,j}$. The sets will be handled by a simple technique shown in [15], that performs optimally.

The data structure required by any constraint $R_{i,j}$ includes a $k \times k$ matrix of pointers, denoted $R_{i,j}$ itself, and a k-cardinality array $C_{i,j}$ of pointers. The generic component $C_{i,j}(w)$ is a pointer to a double linked list containing all the values $x \in D_i$ such that $(x, w) \in R_{i,j}$.

The definition of matrix $R_{i,j}$ is the following:

$$R_{i,j}(v, w) \begin{cases} \text{points to the item } v \text{ in the list } C_{i,j}(w) \text{ if the pair } (v, w) \in R_{i,j} \\ \text{nil} \hspace{4cm} \text{otherwise.} \end{cases}$$

Finally we maintain another array $Count_{i,j}$ where the generic component $Count_{i,j}(v)$ represents the number of elements that are not **nil** in the v-th row of $R_{i,j}$.

The algorithms. Clearly the entire data structure can be initialized in $O(nk^2)$ time. As we remarked above, when we delete a critical pair (a, b) from $R_{i,j}$, we must delete from any constraint $R_{h,i}$ all the pairs of the kind $(*, a)$. In fact, for lemma 8, the value a must be deleted from domain D_i, and then those pairs are uninfluent for maintaining the directional consistency for the arc (v_i, v_j), and for the entire network. This corresponds, in practice, to empty the entire list pointed by $C_{h,i}(a)$ in the constraint $R_{h,i}$.

Our idea is to handle both *explicit* pair deletion, requested by the user, and *implicit* pair deletions, performed as a consequence of an explicit deletion in order to restore consistency. Therefore the algorithm to support pair deletion is implemented by two procedures whose behavior is summarized in the following:

1. Procedure Expl-del($x, y, R_{i,j}$) performs explicit pair deletions, and handles in constant time the deletion of a pair previously subject to an implicit deletion, i.e., a pair such that $R_{i,j}(x, y)$ =nil. Otherwise, if $R_{i,j}(x, y) \neq$nil, it calls procedure Impl-del with the same parameters[5].

2. Procedure Impl-del($x, y, R_{i,j}$) is in charge to delete pairs from the data structure. The deletions of noncritical pairs is performed in constant time, while a critical deletion requires recursive calls in order to restore arc consistency.

The two procedures are shown in figures 1 and 2, and a more detailed description is given in the following.

      Procedure Expl-del($x, y, R_{i,j}$);
1. **begin**
2.   **if** $R_{i,j}(x, y) \neq$nil {the pair (x,y) was never deleted}
3.   **then** Impl-del($x, y, R_{i,j}$);
4. **end.**

Fig. 1. procedure Expl-Del

    The deletion of a pair (a, b) from a binary constraint $R_{i,j}$ is obtained by a call Expl-del($a, b, R_{i,j}$) which, in turn, makes a call to Impl-del($a, b, R_{i,j}$) if the pair (a, b) was not previously deleted (explicitly or implicitly). Procedure Expl-del initially deletes the item a pointed to by $R_{i,j}(a, b)$ in the list appended to $C_{i,j}(b)$, sets $R_{i,j}(a, b)$ to nil and decreases $Count_{i,j}(a)$ to record the deletion.

    Now the algorithm must verifies if the deletion modifies d-arc consistency for the edge (v_i, v_j), i.e., if the pair (a, b) was critical for constraint $R_{i,j}$. This can be done easily by testing whether $Count_{i,j}(a)$ is 0 or not. If $Count_{i,j}(a) \neq 0$ the property is maintained, otherwise for lemma 8 the value a is deleted from D_i setting $R_i(a) = 0$. Each of the operations described above requires $O(1)$ total time to be executed. If the domain D_i becomes empty ($d_i = 0$ after the deletion), the problem becomes unsatisfiable, otherwise the algorithm must try to restore d-arc consistency for any edge (v_h, v_i), if possible, by recursive calls to procedure Impl-del to delete all the pairs $(*, a)$ from any constraint.

    We can observe that the algorithm terminates, because the only case in which this could not be true is when it enters in the nested loop to propagate the previous updates (lines 10–12). In such a case the maximum total number of recursive calls to the procedure Impl-del is $O(nk^2)$ (i.e. the total number of pairs in the entire structure), that is finite.

    The following theorem proves the correctness of the procedures.

---

[5] Marking an element which has been explicitly deleted can be useful to handle intermixed sequences of pair deletions and insertions. Here we neglect this requirement.

Procedure Impl-del($x,y,R_{i,j}$);
1. **begin**
2.   **Delete** x **from** $C_{i,j}(y)$;
3. $R_{i,j}(x,y) :=$ nil;
4.   **decrement** $Count_{i,j}(x)$;
5.   **if** $Count_{i,j}(x) = 0$
6.   **then begin** {a critical deletion occurs}
7.   **Remove** x **from** D_i;
8.   **if** $D_i = \emptyset$
9.   **then** report(''CSP unsatisfiable'')
10.   **else for each edge** (v_h, v_i)
11.   **for each** $z \in C_{h,i}(x)$
                           {i.e., for any pair using the deleted value x in any constraint}
12.   **do** Impl-del($z,x,R_{h,i}$);
13.   **end** {d-arc consistency restored}
14.   **else return**; {noncritical deletion}
15. **end**.

Fig. 2. procedure Impl-Del

Theorem 9. *Given a binary CSP of width 1 which is d-arc consistent, the execution of an* Expl-del($a, b, R_{i,j}$) *to delete any pair* (a, b) *from a constraint* $R_{i,j}$ *restores the property of directional consistency of the whole structure.*

Proof. Only a critical deletion may possibly destroy directional consistency of the network. In such a case this property is guaranteed for the edge (v_i, v_j) in one of the following ways:

1. If the pair (a, b) is not critical for the constraint $R_{i,j}$ for the value $a \in D_i$, then we know that its deletion cannot modify this property.
2. Otherwise, if the pair (a, b) is critical for the constraint $R_{i,j}$, then for lemma 8 its deletion determines the elimination of the value a from domain D_i to restore the property.

In the second case the deletion could modify the consistency of some edge (v_h, v_i). The consistency of this edge is however restored by the deletion of all the pairs of the kind $(*, a)$ from the matrix $R_{h,i}$, by calling the procedure Impl-del. For each of these pairs we can reapply the reasoning used above.

Since in this way we separately restore, or however guarantee, the consistency of each edge in the constraint network, and the network is acyclic, the theorem is proved.

We have already given a worst-case analysis for the execution of a single pair deletion, now we want show that the data structure and the algorithm proposed are efficient in amortized sense. For this kind of analysis we will use the *credit technique* proposed by Tarjan in the seminal paper [17].

In particular we consider an arbitrary sequence of $O(nk^2)$ explicit deletions, i.e., calls to the procedure Expl-del. Now we can prove the following result.

Theorem 10. *The total time required to maintain d-arc consistency of a width-1 CSP, under an arbitrary sequence of at most nk^2 pair deletions, is $O(nk^2)$ on the whole sequence, i.e., $O(1)$ amortized time for each deletion.*

Proof. If we allocate two credits for any pair in our data structure, (i.e., twice the number of items in the double linked lists), these credits are enough to pay for the computational cost of any sequence of pair deletions. In this way we can use at most two credits for each pair deletion.

Each call to procedure Expl-del is paid for by one of the two credits allocated on the delete item. Surely, at least the first call to the procedure Expl-del in the sequence causes a call to an Impl-del with the same parameters to really perform the deletion. So, if we indicate with D the number of calls to the procedure Impl-del directly executed by the Expl-del in the sequence, we have $D \geq 1$.

The credit allocated for the execution of an Impl-del is surely sufficient to execute all the operations preceding the nested loop at the end of the procedure (lines 10–12). So, if we are in the case 1 of theorem 9, the two credit allocated are sufficient for executing the deletion because the algorithm does not enter in the loop. On the other hand, the procedure Impl-del, directly called by the Expl-del, can cause, due to the loop, an arbitrary sequence of recursive calls to itself. Each of these calls performs a real deletion of an item from our data structure. So, if $t_j \leq nk^2$ is the number of recursive calls that procedure Impl-del, directly called by the j-th Expl-del in the sequence, performs in cascade, then the total number T of real deletions, undirectly executed, on the whole sequence is:

$$T = \sum_{j=1}^{nk^2} t_j \leq nk^2.$$

The equality $T = nk^2$ can never verify because we know that $D \geq 1$. Finally we have:

$$T = \sum_{j=1}^{nk^2} t_j < nk^2.$$

The total number of credits allocated to perform the T deletions described above is $2 * T < 2nk^2$. But only the half of them is really used, one for each of the T undirect calls to the procedure Impl-del.

So during the whole sequence of pair deletions we have T saved credits, each one of them is however used as follows. To each one of the T undirect calls to the procedure Impl-del to delete a pair (a, b) from an arbitrary constraint $R_{i,j}$, corresponds to a possible future call of the kind Expl-del$(a, b, R_{i,j})$, which will find $R_{i,j}(a, b) =$ nil. In such a case the execution of this call does determines no further call to the Impl-del, so one credit is sufficient to complete it. To this aim we can use the credit saved during the execution of the Impl-del$(a, b, R_{i,j})$ called previously.

If we extend this reasoning to each of the T undirect calls to the procedure Impl-del on the sequence, we have that each of the credit saved during the execution is used later.

The considerations above proves that the total number of credits necessary to complete an arbitrary sequence of at most nk^2 pair deletions is:

$$\sum_{j=1}^{nk^2} 2 = 2nk^2 = O(nk^2)$$

Hence $O(1)$ amortized time is sufficient to perform each deletion.

3.2   Arc (constraint) insertions

In this section we propose an extension of the data structure used in the previous section, and a dynamic algorithm that allow us to maintain a solution to an acyclic binary CSP in $O(nk^2 + kn^2)$ total time, under an arbitrary sequence of arc insertions, starting with an empty constraint graph (a forest), in which all the constraints are universal constraints.

A generic insertion of an arc (x_i, x_j) is treated here as a sequence of deletions from the matrix $R_{i,j}$ of all the pairs that are not allowed after the insertion.

To ensure a backtrack-free search for the solution for a tree-like structure under a sequence of arc insertions, it is not sufficient to maintain the property of d-arc consistency, but, for theorem 5 it is necessary to maintain also the property of width 1. Hence we are required to verify that the constraint graph obtained after each insertion is a forest.

Lemma 11. *Given a forest F, the insertion of an arc (x_i, x_j) maintains F with width 1 only if the nodes x_i and x_j belong to different trees in the forest F.*

Proof. The thesis immediately follows from the consideration that a graph has width 1 iff it is acyclic.

The data structure.   To achieve our aim, i.e., maintaining d-arc consistency and width 1 for acyclic binary CSP under an arbitrary sequence of arc insertions, we use an extension of the data structure described in section 3.1. Precisely we use the following additional data structures:

- A double linked list L of pointers to the roots of the trees in the forest.
- A type-record data structure to represent the nodes in the forest. Each item of this kind contains the following fields: a pointer p to its parent in the forest; a pointer child to the list of its children in the corresponding tree; a pointer list, which is significant only for the root nodes, to the item in L that points to it.

Using this extended data structure we can maintain efficiently the property of d-arc consistency in the same way described in section 3.1, and the property of width 1 as described in the next section.

The algorithm. To insert an edge (x_i, x_j) our algorithm performs the following operations: the paths from the nodes x_i and x_j to the respective roots are crossed backward, and verifies whether these roots are equal. In such a case the algorithm halts because the width of the actual constraint graph is 2. One of the two trees involved in the insertion is randomly chosen, say $root(x_i)$, and "reversed" making x_i its new root, properly updating the additional structures involved. The directional consistency of the edges involved in this process is suitably restored, to maintain this property on the whole network. $p(x_i)$ is set to point to x_j and the item in the double linked list L which points to x_i is deleted. The universal constraint $R_{i,j}$ is created, and a sequence of deletions of the pairs that are not allowed is performed. These updates are propagated on the structure to all the edges in the path from x_j to $root(x_j)$.

Our algorithm is structured in form of three procedures denoted Insert, Update and Reverse. If (x_i, x_j) is the edge which must be inserted, we indicate with $r(x_i)$ the root of the corresponding tree having size denoted as $|r(x_i)|$, and with $p(x_i)$ the parent of the node x_i in this tree. With these conventions the procedures are organized as follows.

```
      Procedure Insert(xᵢ, xⱼ : node; Rᵢ,ⱼ:constraint);
  1. begin
  2.     if r(xᵢ) ≠ r(xⱼ)
  3.         then begin
  4.             Update(xᵢ, xⱼ, Rᵢ,ⱼ);
  5.             Delete the item in L which points to xᵢ;
  6.         end
  7.     else report(''width(G) > 1'')
  8. end.
```

Fig. 3. procedure Insert

Procedure Insert (in figure 3) checks in $O(|r(x_i)| + |r(x_j)|)$ time (crossing backward the paths from x_i and x_j to $r(x_i)$ and $r(x_j)$ respectively) whether $r(x_i) \neq r(x_j)$[6]. In such a case the insertion is allowed and the algorithm properly call the procedure Update described in figure 4. In the other case the algorithm halts because the insertion performed introduces a cycle in the actual constraint graph.

In the procedures the union of the domains of all the variables is denoted as D. Procedure Update calls procedure Reverse only if the node x_i is not a root. Then it inserts the edge required in $O(1)$ time pointing $p(x_i)$ to x_j and inserting in $O(1)$

---

[6] These operations are not critically related to the complexity of our algorithms. In other words we are not interested to check acyclicity by using a more sophisticated approach (say, by some *set-union-find* algorithm), since we have to perform specific operations on any crossed arc.

```
    Procedure Update(x_i, x_j: node; R: constraint);
1.  begin
2.      if x_i ≠ r(x_i)
3.          then Reverse(x_i, p(x_i), p(p(x_i)));
4.      p(x_i) := x_j;
5.      R_{j,i} := R;
6.      insert the item x_i in the list child(x_j);
7.      for each a ∈ D
8.          do if (a ∉ D_i)
9.              then for each z ∈ C_{j,i}(a)
10.                 do Impl-del(z, a, R_{j,i});
11. end.
```

Fig. 4. procedure Update

time x_i in the list of the children of x_j. Then it creates in $O(k^2)$ time the universal matrix $R_{i,j}$, and propagates the possible updates due to the insertion to all the edges on the path between x_j and $r(x_j)$ in $O(|r(x_j)|k^2)$ total time. Procedure **Reverse** is described in figure 5.

```
    Procedure Reverse(x_i, x_j, x_h : node);
1.  begin
2.      if x_j ≠ x_h then Reverse(x_j, x_h, p(x_h));
3.      p(x_j) := x_i;
4.      p(x_i) := x_i;
5.      Update the lists child(x_i) e child(x_j);
6.      for each a ∈ D
7.          do if a ∉ D_j
8.              then for each z ∈ C_{i,j} do Impl-del(z, a, R_{i,j});
9.  end.
```

Fig. 5. procedure Reverse

This procedure reverses the direction of all the edges on the path from x_i to $r(x_i)$, and for each one of them ensure the consistency in this new direction in $O(|r(x_i)|k^2)$ total time.

By the considerations above we have that the total time necessary, in the worst case, to insert an edge between two trees T_1 and T_2 in a forest is $O(k^2(|T_1|+|T_2|+1))$. Furthermore this also shows as the choice of the tree to "reverse", after each edge insertion, is completely uninfluent.

The following theorem proves the correctness of the proposed procedures.

Theorem 12. *Let G be the constraint graph of a width-1 binary CSP which is d-arc consistent. The attempt to insert a generic edge (x_i, x_j) with constraint $R_{i,j}$ executing an Insert$(x_i, x_j, R_{i,j})$ determines one of the following results:*

1. *The insertion is not performed if it introduces a cycle in G;*
2. *If the insertion does not introduce cycles the property of d-arc consistency of the new constraint graph G' is guaranteed.*

Proof. To prove the acyclicity of the new graph G' we must show that, after any call to Insert$(x_i, x_j, R_{i,j})$ on a forest G, G' is still a forest.

Procedure Insert verifies that the nodes x_i and x_j have different roots (in the opposite case the insertion is not performed because case 1 occurs) and calls procedure Update. It calls procedure Reverse which traverses the path from node x_i to $r(x_i)$ changing the direction of all the edges on this path. Finally the algorithm traverses the path from the old root of node x_i to $r(x_j)$ restoring the directional consistency of all the edges encountered using procedure Impl-del described in section 3.1.

To maintain the directional consistency of the whole structure, the same process applied for pair deletion is used, exploiting the ordering given by the tree G' obtained after each insertion.

Therefore the maintenance of the d-arc consistency for the whole structure after each insertion is guaranteed by the same motivations given in the proof of theorem 9.

Now we will prove that the proposed data structures and algorithms are efficient in amortized sense. We will prove in particular that the total time necessary to execute an arbitrary sequence of at most $O(n)$ edge insertions, each one of $O(k^2)$ size, is $O(nk(n+k))$, instead of $O(n^2k^2)$ which we would obtain executing each time the off-line algorithm Revise [10]. Finally we can prove the following result.

Theorem 13. *The total time required to maintain the properties of d-arc consistency and width 1 for a binary CSP, under an arbitrary sequence of $O(n)$ edge insertions, each one of size $O(k^2)$, starting from an empty constraint graph and leaving the graph acyclic, is $O(nk(n+k))$.*

Proof. To maintain the property of d-arc consistency on the whole structure under the sequence of $O(nk^2)$ pair deletions, corresponding to the $O(n)$ edge insertion in the sequence, the algorithm uses the same technique described in section 3.1. This technique allows us to perform each pair deletion in $O(1)$ amortized time on the whole sequence, and to propagate the possible updates due to the deletion upwards on the actual ordering. The total time necessary to maintain this property under the whole edge insertion sequence is $O(nk^2)$.

To enforce width 1 for the constraint graph, our algorithm must perform in the worst case, for each edge insertion, a Reverse for one of the two trees involved. We must verify what is the total time necessary to perform the $O(n)$ calls to the procedure Reverse on the sequence. We know that, in the worst case, the procedure Reverse performs $O(n^2)$ edge traversal. This is the case in which the constraint graph is a chain $< 1, 2, \ldots, n >$ and the edges are inserted in the order $< (1,3), (2,4), \ldots, (n-3, n-1), (n-2, n) >$.

During each traversing of an edge the procedure **Reverse** perform some of the $O(nk^2)$ possible pair deletions. Furthermore, for each traversing the procedure **Reverse** perform the cycle starting at line 6 which requires $O(k)$ time to be executed. So the total time necessary to maintain the acyclicity of the constraint graph is $O(kn^2)$.

Finally the total time required to perform an arbitrary sequence of $O(n)$ edge insertions, each one of $O(k^2)$ size, is $O(kn^2 + nk^2) = O(nk(n+k))$ which corresponds to have, for each $O(k^2)$ sized edge insertion in the sequence, the amortized cost $O(\max\{n/k, 1\})$.

We know that inserting an edge and deleting a pair from a tree-like CSP are homogeneous operations. So we can perform on the same constraint graph sequences of such operations using the same data structure described in section 3.2.

The next theorem result follows directly from theorems 9 and 12.

Theorem 14. *The total time necessary to maintain the properties of d-arc consistency and width 1 for a binary CSP under a sequence of at most $O(n)$ edge insertions and $O(nk^2)$ pair deletions, starting from an empty constraint graph is $O(nk(n+k))$.*

Before concluding this section, we show as alternating insertion and deletion of tuples from constraints by using our data structure is also possible, but the time required to deal with this case can be as much as $O(nk^2)$ for any subsequence of pair deletions (while insertions are performed in constant time). This can be handled by performing *lazy* insertions. This approach requires an operation that we call *restore*, consisting in applying the off-line $O(nk^2)$ algorithm to recompute from scratch the content of data structures. Some details follow:

- when an explicit deletion occurs, the corresponding entry in the binary matrix representation is *marked*: in this way both the original constraints (the unmarked items) and the pruned version updated by the algorithm (the non-null items) are represented in the data structures;
- when the problem becomes unsatisfiable, and at least one insertion has been performed since the last *restore* occurred, a new *restore* is performed again.

In this way the newly introduced values are considered only when the old ones are not sufficient to build a solution for the considered CSP. Since the time required by deletions between two subsequent *restore*'s is at most $O(nk^2)$, computing a *restore* does not modify the asymptotic performance of our algorithm.

4 Conclusions

In this paper we deal with the constraint satisfaction problem in a dynamic framework. In particular we consider a simple case, where the constraint are binary and the network is acyclic. Following an approach proposed by [5] for the static case, we propose algorithms supporting deletion of allowed tuples from constraints, and insertion of new constraints, while maintaining a data structure implicitly representing all possible solution for the considered CSP.

It is easy to extend the class of constraints suitable to be tackled by these algorithm to include all (i.e. not only binary) *acyclic CSP*. In particular this is the class of those CSP for which the constraint hypergraph is *strongly acyclic*.

More precisely given an undirect hypergraph $H =< N, A >$, its *FD-graph* is a bipartite undirected graph $G(H) =< N_H, A_H >$, where $N_H = N \cup A$ and $A_H = \{(x, y) \mid x \in N, y \in A, \text{ and } x \in y\}$. An undirected hypergraph $H =< N, A >$ is *strongly acyclic* if the corresponding FD-graph is a tree. A CSP is *acyclic* if the corresponding labeled constraint hypergraph $H_l =< N, A, l >$ is strongly acyclic.

Given a generic labeled constraint hypergraph $H_l =< N, A, l >$ it is easy to built up the corresponding labeled FD-graph $G(H_l) =< N_{H_l}, A_{H_l}, l' >$, and, if H_l is *strongly acyclic*, this process require $O(nk^2)$ time, where:

- $n = |N|$ i.e. n is equal to the number of variables in the considered CSP (note that $|N_H| \leq 2n$);
- $k = \max_{a_j \in A}\{|e_j|\}$, i.e., the cardinality of the constraints in the original CSP (note that for any variable v_i, $|D_i| \leq k$).

If the FD-graph $G(H_l)$ is acyclic, all the algorithms described in the previous sections can be simply extended to $G(H_l)$ and so to H_l. The following result holds.

Theorem 15. *Let us consider an arbitrary CSP defined on n variables and using arbitrary constraints with maximum size k. If the considered CSP is acyclic, then there exist data structures and algorithms supporting arbitrary sequences of operations of the following kinds (provided that the resulting CSP remains acyclic):*

1. *deletion of k-tuples from any constraint;*
2. *insertion of new arbitrary constraints.*

The total time required to maintain a solution for such a dynamic CSP is $O(nk(n + k))$. If the allowed operations are restricted to be tuple deletions, the required total time is $O(nk^2)$, that is $O(1)$ per update in any sequence of $\Theta(nk^2)$ pair deletions. In both cases, reporting a solution for the current CSP, i.e. an assignment for the n variables, requires $O(n)$ time.

Acknowledgements

We like to thank Claudio Arbib for useful suggestions and comments.

References

1. A. V. Aho, J. E. Hopcroft, J. D. Ullman. *The design and analysis of computer algorithms.* Addison-Wesley, Reading, MA, 1974.
2. G. Ausiello, A. D'Atri, D. Saccà. Minimal representation of directed Hypergraphs. *SIAM J. Comput.*, 15:418-431, 1986.
3. C. Berge. *Graphs and Hypergraphs.* North Holland, Amsterdam, 1973.
4. R. Detcher. Enhancement schemes for constraint processing: Backjumping, Learning and Cutset decomposition. *Artificial Intelligence*, 41, 1989.

5. R. Detcher, J. Pearl. Network based heuristic for constraint satisfaction problems. *Artificial Intelligence*, 34, 1988.
6. E. C. Freuder. Synthesizing constraint expression. *Commun. ACM*, 21, 11, 1978.
7. E. C. Freuder. A sufficient condition for Backtrack-free search. *J. ACM*, 29, 1, 1982.
8. E. C. Freuder. A sufficient condition for Backtrack-bounded search. *J. ACM*, 32, 4, 1985.
9. E. C. Freuder, A. K. Mackworth. The complexity of some polynomial network consistency algorithms for Constraint Satisfaction Problems. *Artificial Intelligence*, 25, 1985.
10. A. K. Mackworth. Consistency in networks of relations. *Artificial Intelligence*, 8, 1977.
11. U. Montanari. Network of constraints: fundamental properties and application to picture processing. *Information Science*, 7:95–132, 1974.
12. U. Montanari, F. Rossi. An efficient algorithm for the solution of hierarchical networks of constraints. *Lecture Notes in Computer Science*, 291, Springer Verlag, Berlin, 1986.
13. U. Montanari, F. Rossi. Fundamental properties of networks of constraints: a new formulation. In: L. Kanal and V. Kumar, eds., *Search in Artificial Intelligence*, Springer Verlag, Berlin, 426–449, 1988.
14. U. Montanari, F. Rossi. Constraint relaxation may be perfect, *Artificial Intelligence*, 48, 1991.
15. U. Nanni, P. Terrevoli. A fully dynamic data structure for path expressions on dags. *R.A.I.R.O. Theoretical Informatics and Applications*, to appear. Technical Report *ESPRIT-ALCOM*.
16. R. E. Tarjan. *Data structures and network algorithms*, volume 44 of *CBMS-NSF Regional Conference Series in Applied Mathematics*. SIAM, 1983.
17. R. E. Tarjan. Amortized computational complexity. *SIAM J. Alg. Disc. Meth.*, 6:306–318, 1985.

Efficient Reorganization of Binary Search Trees

Micha Hofri [*]
Department of Computer Science
The University of Houston, Houston Tx 77204-3475, USA

Hadas Shachnai[**]
Computer Science Department
Technion, Israel Institute of Technology, Haifa 32000, Israel

Abstract. We consider the problem of maintaining a binary search tree that minimizes the average access cost with respect to randomly generated requests. We analyze scenarios, in which the accesses are generated according to a vector of probabilities, which is fixed but unknown.

In this paper we devise policies for modifying the tree structure dynamically, using rotations of accessed elements towards the root. Our aim is to produce good approximations of the optimal order of the tree, while minimizing the amount of rotations.

We first introduce the *Move Once* (MO) rule, under which the average access cost to the tree is shown to equal the average access cost under the commonly used *Move to the Root* (MTR), at each reference. The advantage of MO over other rules is that MO relocates each of the items in the tree at most once. Consequently, modifying the tree by the MO rule results in $O(n\lg n)$ rotations (with n the number of items) for any *infinite* sequence of accesses.

Then we propose to combine the MO with the usage of counters (accumulating the reference history for each item), that provide approximations of the reference probabilities. We show, that for any $\delta, \alpha > 0$, this rule (which we call MOUCS) approaches the optimal cost to within a difference of δ with probability higher than $1 - \alpha$, after a number of accesses, which is linear in n times $1/\alpha$ times $1/\delta^2$.

1 Introduction

The *Binary Search Tree* (BST) is commonly used for storing tables and lists. The advantage of a tree is that it allows an efficient search of the table (or list). Typically, the search is most efficient when the tree is kept as balanced as possible, and when frequently accessed elements are close to the root. We study heuristics which maintain a BST in a nearly optimal form.

[*] e-mail: hofri@cs.uh.edu.
[**] Currently at IBM T.J. Watson Research Center, P.O. Box 704, Yorktown Heights, NY 10598. e-mail: hadas@watson.ibm.com. Author supported in part by the Technion V.P.R. Fund – E. and J. Bishop Research Fund and by the Fund for the Promotion of Research at the Technion.

The scenario considered has a set of n records in random storage, $L = \{R_1, \ldots, R_n\}$. The record R_i is uniquely identified by the key K_i, for $1 \leq i \leq n$, and the keys satisfy a total order. The set is maintained as a BST. The records are accessed according to a multinomial distribution driven by the fixed *Reference Probability Vector* (RPV): $\bar{p} = (p_1, \ldots, p_n)$. Thus, R_i may be requested at any stage with the same probability p_i, independently of previous requests and the state of the tree – and in particular of the location of this R_i. This is the so-called *independent reference model* (IRM). Since the RPV and L are constant, the passage of time is realized by the sequence of references. There is no other notion of time in the model.

Each reference requires a search for a record in the tree. The cost of a single access is defined as the number of key comparisons needed to locate the specified record.
The order by which the records are initially inserted into the tree is assumed to be random (with equal probability over all possible permutations). Different initial insertion sequences may result in different trees, usually with different expected access cost.

The access probabilities listed in the RPV $\bar{p}$ are assumed unknown. Were they known, we could restructure the tree, using a dynamic programming approach to provide the smallest possible expected cost. Since the RPV is constant, so would be the optimal structure. With $\bar{p}$ unknown, we are reduced to looking at policies that use the accumulating reference history to adapt the tree structure with the goal of rearranging the records so that the expected access cost is minimized.

The reorganization process incurs a cost as well: the manipulations performed on the tree when its structure is modified. The only operations used for this are *rotations*, operations that exchange the 'rotated' node with its parent, while maintaining the key-order in the tree. Figure 1 shows the tree modifications that result. Note that the inverse of the rotation operation is a rotation as well. The cost of the reorganization is defined as the number of rotations, since each rotation requires essentially the same computing time.

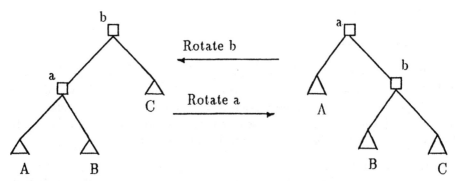

Fig. 1. Single left-child and right-child rotations that reflect $(A) < K_a < (B) < K_b < (C)$

The two cost components of key comparisons and rotations are denoted by C and R respectively. A few performance measures which are of interest in this context are:

1. The access cost following the mth reference (and possible reorganization), $m \geq 0$.
2. The asymptotic access cost, especially its expected value.
3. The work done in reorganizing the tree, such as the moments of the total number of rotations.

In addition to the limiting cost of C, it is also interesting to consider the rate at which it approaches this limit: since most data processing systems exist for a finite time only, a policy which reduces the access cost promptly may be preferable to one that does it more slowly, even if the limiting cost of the latter is somewhat lower.

The problem of reorganizing a BST to minimize the average access cost has been widely studied. However, most of the works focus either on the asymptotic value of C, or an "amortized" analysis of the cost, which combines C and R, but is limited to considering a worst-case scenario [8],[19] [16], and typically yields only bounds, at that. Such an analysis cannot assess the advantage of using properties of the IRM (while—since it considers the worst case—including scenarios where this assumption does not hold). The results in [4] refer to the case when the elements of the RPV are known only up to a permutation.

Some of the research focused on the situation where the RPV is *known*, with the goal of finding the optimal tree – or achieving a nearly optimal one, with a smaller computational effort. An early outline is given in [15]. A survey of recent work on balanced BSTs appears in [18]. Recently, parallel algorithms have been considered for construction of optimal and nearly optimal BSTs ([2],[13]).

In this paper we devise and analyze heuristics which achieve a close approximation to the optimal BST, with lower organization costs than any of the previously studied heuristics.

Section 2 defines some additional notation, and presents the dynamic programming algorithm that constructs the optimal tree for a known RPV.
In Section 3 we discuss the *Move Once* (MO) rule, which achieves the same C cost as the *Move to the Root* (MTR) rule ([1]), but requires at most $n - 1$ reorganization steps for any reference sequence to the tree.
We then propose in section 4 a method for approximating the optimal search tree, which improves the asymptotic average cost obtained by the MTR. Our method, which we call *Move Once and Use Counter Scheme* (MOUCS) guarantees, that for any δ and $\alpha > 0$, the tree is dynamically reorganized until the average access cost is within δ of the optimal cost, with probability of at least $1 - \alpha$. We obtain a distribution-free bound on the running time of the algorithm, which is linear in n (the size of the tree) times $1/\alpha$ times $1/\delta^2$.

2 Preliminaries

Let $C(T_n)$ denote the average access cost to a BST T of n elements, with the access probabilities $p_1, \ldots, p_n$, then

$$C(T_n) = 1 + \sum_{i=1}^{n} p_i \cdot (\text{level}\,(R_i)). \tag{1}$$

Under the IRM, for any set of keys with a given RPV, there exists an optimal *static* BST. We denote by $C(\text{OPT}|\bar{p})$ the average access cost in such an optimal tree.

The optimal tree structure and its expected cost are straightforward to compute using the following Dynamic Programming equations, which need to be satisfied at every internal node (adapted from [15]). Let $C(i,j)$ be the expected access cost of the optimal subtree that consists of records $R_{i+1}, R_{i+2}, \ldots, R_j$. Then $C(0,n)$ is the $C(\text{OPT}|\bar{p})$ defined above. We also define $\pi_{i,j} = \sum_{k=i}^{j} p_k$.

$$C(i,i) = 0,$$
$$C(i,j) = \pi_{i+1,j} + \min_{i < k \leq j}(C(i,k-1) + C(k,j)) \quad \text{for } 0 \leq i < j \leq n, \tag{2}$$

When the access probabilities are *unknown*, a dynamic reorganization of the tree may be used to achieve an approximation of the optimal order. Some of the well known modification rules are studied in [1] and [4]. Various performance measures were considered for this model. With a given reorganization policy B and an unknown RPV $\bar{p}$, the following costs are used below:

1. The average access cost after the mth reference, $m \geq 0$, denoted by $C_m(\text{B}|\bar{p})$, where

$$C_m(B|\bar{p}) = 1 + \sum_{i=1}^{n} p_i \cdot E[\,\text{Level}\,(i) \mid B, \bar{p}\,], \tag{3}$$

   Note that the expected level at which an item may be found, under a policy B, is determined by the RPV and the initial state of the tree, possibly as a result of the order the elements were inserted into the tree. As a rule we average over all possible insertion sequences, considering them equiprobable.[3] Under certain policies, the initial state becomes irrelevant following a large number of references and the changes in the tree they trigger. In particular, this must be a property of any reorganization policy which approximates the optimal tree after a finite (possibly long) sequence of searches.
2. The expected access cost in the limiting state:

$$C(B|\bar{p}) = \lim_{m \to \infty} C_m(B|\bar{p}).$$

---

[3] This does *not* translate to a uniform distribution over initial tree states, since the number of sequences that result in a given tree state is not the same for all tree shapes.

3. The total expected number of rotations induced by an input string, characterized either by its size (e.g. m requests) or by the number of distinct records it references. In particular – a sequence that contains each record at least once.

3 The Move Once (MO) Rule

3.1 The Average Cost of a Single Access

Allan and Munro [1] analyzed the Move To the Root (MTR) rule in detail, assuming the IRM. This rule is the counterpart of the Move To the Front (MTF) rule for linear lists: a referenced record is rotated to the root of the tree (unless it is there already). They showed an upper bound on the ratio $C'(\text{MTR}|\bar{p})/C(\text{OPT}|\bar{p})$ for any distribution (their Thm. 3.3), and also estimated its rate of convergence (their Thm. 5.1).

The MTR rule is on the one hand more attractive than the MTF for a linear list, since the limiting value of its access cost can be shown to be closer to the optimal cost; on the other hand, the R component of its cost is even more pronounced than with a list, where any rearrangement uses the same time; here, moving a record to the root uses the same number of rotations as the number of steps done to reach the record. Hence it makes sense to look for rules that use less expensive modifications. In [10] it was shown that for a linear list, moving a record at most once (when it is first referenced) to the tail of the sublist of records that were moved before, achieves the same expected cost as the MTF, at any finite time. We propose to use the same principle for reorganizing BSTs – a record is only moved the first time it is referenced. It is then rotated towards the root, until its parent is a record that has already been referenced. The first referenced record goes of course all the way to the top (Move Once, MO).

Allan and Munro [1] consider a similar rule, calling it the First Request Rule, and show it has the same *asymptotic* cost as the MTR. However, for BSTs, just as for linear lists, more can be claimed:

Theorem 1. *Let a BST be referenced according to the IRM with the RPV $\bar{p}$. The rules MTR and MO have the same expected access costs for the mth request, $m \geq 1$.*

Proof: Both rules start with the same tree (or with trees selected at random using the same initial distribution, which amounts to the same thing). The key to the equality of their expected access costs is the following claim:

For a given initial BST, let $T_B(I)$ be the BST resulting from processing the reference string I with the reorganization rule B. Then

$$T_{MTR}(I) = T_{MO}(I^R),$$

where the string I^R is the reverse of I.
We give the proof of the claim in the Appendix.

Under the IRM, both strings have precisely the same probability, hence the expected access costs of the two rules are equal. It may seem surprising, since usually they construct for the same input string entirely different trees. The important difference is that MO uses far fewer rotations than MTR, and moreover, the latter churns its tree indefinitely, whereas MO rests after a time which has a finite expectation. ∎

Hence we can use the results in [1] for the average cost under MTR to state the following theorem.

Theorem 2. [1] *The expected access cost to a* BST *reorganized with the* MO *policy, after m references, is given by*

$$C_m(\text{MO}|\bar{p}) = 1 + \sum_{1 \le i < j \le n} \left[\frac{2p_i p_j}{\pi_{i,j}} + (1 - \pi_{i,j})^m \left(\frac{p_i + p_j}{j - i + 1} - \frac{2p_i p_j}{\pi_{i,j}} \right) \right] \qquad (4)$$

where $\pi_{i,j} = \sum_{k=i}^{j} p_k$.

Note: This result applies when the tree is considered to have been formed by a uniformly distributed insertion sequence. If it were created by the same applications which gave rise to the RPV we used above, then its initial form has the asymptotic distribution induced by the MO policy. This has then the limiting expected value

$$C(\text{MO}|\bar{p}) = 1 + \sum_{1 \le i < j \le n} \frac{2p_i p_j}{\pi_{i,j}}. \qquad (5)$$

3.2   The Expected Number of Rotations

We consider R_n, the total number of rotations the MO policy requires to organize a BST of size n (that is created by a random sequence) till all records have been accessed at least once.

The size of such a sequence is known as the length of the Coupon Collector Search. It is discussed in detail in [5]. It depends critically on the RPV; its shortest expected value is obtained for uniform $\bar{p}$ and equals then asymptotically $n \log n$. While the number of rotations for the first few references would be typically approximately $\log n$, we should expect most subsequent references to require few rotations, if any.

The number of rotations, R_n for a tree of size n, depends on two distributions. One is used to generate the insertion sequence that creates the initial tree, and one governs subsequent accesses. It would be more accurate to say it depends on the relation between the two. This general statement has an exception: if the first distribution is uniform – every permutation of the records is equally likely to serve as the insertion sequence – then R_n does *not* depend on the *access* RPV. The reason is apparent from the equation we derive now. Consider the first reference. It addresses some record R_I, where I is the position of that record in the total order of the keys. Since the tree was created with the uniform distribution, then regardless of the values of I and of p_I itself, R_I is in depth D_n (= the depth of a randomly selected node in a randomly constructed BST) – independently of I. Hence we may assume that the variable I is uniformly distributed on $[1, n]$. D_n is also the number of rotations that bring it to the root. Once this is done it will have two subtrees of sizes $I - 1$ and $n - I$, and again, their structure is that of randomly created BSTs. The MO policy translates to independent reorganization of the subtrees; we find then

$$R_n = D_n + R_{I-1} + R_{n-I}. \tag{6}$$

The statistics of D_n are well known; it satisfies a recursion even simpler than (6): $D_n = 1 + D_{I-1} + D_{n-I}$. ¿From this it is easy to obtain its *probability generating function* (PGF): $[n(1 - 2z)]^{-1}(1 - (-1)^n \binom{-2z}{n})$. The expected value is $d_n \equiv E[D_n] = 2(1 + \frac{1}{n})H_n - 4$ (where H_n is the nth harmonic number). Its variance is given by $u_n \equiv V[D_n] = 2(n + 5)H_n/n + 4[1 - (n + 1)H_n^{(2)}/n - (n + 1)H_n^2/n^2]$ (and here $H_n^{(2)}$ is the nth second-order harmonic number).

Now let us return to rotations: Taking expected values (with respect to the entire access sequence used by the MO policy) of equation (6) we derive the immediate difference equation

$$r_{n+1} \equiv E[R_{n+1}] = d_{n+1} - \frac{n}{n + 1}d_n + \frac{n + 2}{n + 1}r_n, \tag{7}$$

which has the solution

$$r_n = 2n(H_n^{(2)} - 1) - 2H_n + 2H_n^{(2)}. \tag{8}$$

For not-too-small n, a good approximation of r_n is given by $r_n \approx n(\pi^2/3 - 2) - 2\log n + 2.3950$.
The total expected number of rotations *per record* in the tree is then less than 1.3.
The PGF of R_n does not appear to be easy to obtain. Even the variance $v_n \equiv V[R_n]$, which satisfies a relation very much like (7), does not seem to have a useful closed form representation. An asymptotic estimate is obtainable, though the explicit form is very complicated. It boils down to

$$v_n \approx 1.165381n - 4\log^2 n - 10.617725\log n + O(1). \tag{9}$$

4 Reference Counters and Approximately Optimal Trees

The glaring difference between the use of reference counters for the reorganization of linear lists and of BSTs is that for the first storage mode the Counter Scheme (CS) – the policy that keeps the records ordered by their counters – converges to the optimal order, while unless tree reorganization is "free," there appears to be no such simple rule for BSTs that results in the optimal structure.

4.1 The Counter Scheme and Dynamic Programming

With linear lists, the CS was shown to be optimal not only asymptotically, but also for every *finite* reference string (see in [11]), and it is tempting to assume such a rule for BSTs. This does not seem to be the case. In particular, the so-called "monotonic BST", which like the CS keeps records with higher counters closer to the root, will usually fail to result in the optimal structure, for the same reason that using a *known* RPV to structure a BST monotonically fails to reproduce the tree which is computed by the Dynamic Programming (DP) algorithm of section 2.

It is also known that the cost of the monotonic tree can be unboundedly higher than that of the optimal one (specifically – their ratio can be as high as roughly $n/\log n$, [17]).

But all is not lost. We can combine the CS with the DP algorithm as follows. The counters provide estimates for the RPV which could be used in DP to produce a tree which is the optimal one for the estimates, but would usually be in fact suboptimal. This computation takes a (deterministic) time of order n^2; this is non-trivial for a large tree. Hence we would like to do it once, and to do it right. This requires

1. That the estimates should be good enough for the deviation from optimality to be tolerable.
2. An efficient management of the counters which add space overhead. As we show below, the total number of references needed is typically a moderate multiple of n. Hence, unless for some reason especially small counting registers are used, we need not worry about their potential overflow.

What the procedure needs is a stopping criterion, a way to determine when the estimates are good enough to stop the MO phase. The criterion must relate the total number of references, possibly with some information about the estimated RPV, and the nearness of the suboptimal value to the optimal one.

We suggest the following compound policy MOUCS (for Move Once and Use Counter Scheme):

Reorganization Rule MOUCS. This rule has two phases:
Phase A: Use the rule MO and also compile reference counters, C_i for R_i, during a total of m_0 references, where we calculate a suitable m_0 below.
Phase B: use $\{C_i/m_0\}$ as estimates for the RPV; compute the "ostensibly optimal" BST using them as input for the DP algorithm, restructure the tree accordingly and stop reorganizing.

For the access cost during Phase A we have an explicit, if cumbersome result in equation (4).
Allen and Munro use equation (4) to show that the MTR rule produces a tree with an expected access cost that differs from its limiting value by at most one, within $\lceil n \log n/c \rceil$ references. [4] They also show, however, that this limiting value,

---

[4] The remarkable fact about this value is that it is far less than the expected number of references before all records are referenced at least once!

$C(\text{MTR}|\bar{p})$, can exceed the optimal one by some 40%. We would like to do better.

In the following result we quantify the efficiency of the MOUCS rule in terms of convergence to the optimal tree vs. the length of the request sequence. Denote by $\hat{C}(\text{OPT}|\bar{p})$ the expected access cost to the tree constructed in Phase B for the estimate $\hat{p}$ to the RPV $\bar{p}$.

Theorem 3. *For any unknown RPV $\bar{p}$, $\delta > 0$ and $\alpha < 1$,*

$$Prob\left(\left|\hat{C}(\text{OPT}|\bar{p}) - C(\text{OPT}|\bar{p})\right| > \delta\right) < \alpha \tag{10}$$

after a sequence of m_0 accesses to the tree, where

$$m_0 = \frac{2.6675n}{\delta^2\alpha}, \tag{11}$$

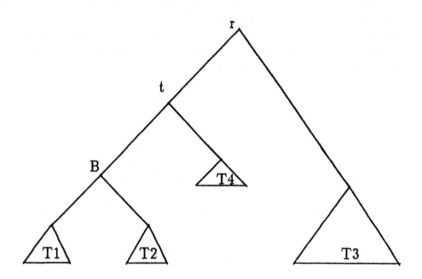

Fig. 2. The optimal tree T

We use in the proof the following Lemmas, which relate weights (=access probabilities) of subtrees to their position in the optimal BST. We remind the reader that the level of a node was defined as its distance from the root.

Lemma 4. *Let T be an optimal BST as given in Figure 2: Node B is at level 2, and P is the weight of T. Let p_B, p_t, p_r denote the weights of the nodes B, t, r respectively, P_t is the weight of the tree rooted at t, and P_B is the weight of the tree rooted at B, and for $1 \leq i \leq 4$, P_i are the weights of the subtrees T_i, then $\forall a \in [0, 1]$*

$$either \quad P_t \leq aP \quad or \quad P_B < (1-a)P. \tag{12}$$

The proof is given in the full paper.

Lemma 2 improves the bound on the maximal level of an item in an optimal BST as given in [9].

Lemma 5. *For any set of weights $p_1, \ldots, p_n$ corresponding to the keys $K_1, \ldots, K_n$, such that*
$S = \sum_{i=1}^{n} p_i \leq 1$, *if L_i is the level of K_i in an optimal BST (where the level of the root is 0), then*

$$p_i \leq \left(\frac{1}{\varphi}\right)^{L_i - 1} . \tag{13}$$

where φ is given by $(1 + \sqrt{5})/2$, the celebrated golden ratio.

Proof: Using Lemma 4 with $a = 1/\varphi = \frac{\sqrt{5}-1}{2}$, the proof is similar to the proof of Lemma 2 in [17]. In addition we get an *a-fortiori* bound by replacing the weight of a subtree by that of its root. ∎

Proof of Theorem 3: Let $\hat{p}$ be the estimate for $\bar{p}$, obtained after a sequence of m references, and let $\hat{l}_i, l_i$ be the levels of R_i in the optimal trees for $\hat{p}, \bar{p}$ respectively, then using the optimality of $\{\hat{l}_i\}$ for $\{\hat{p}_i\}$,

$$\left| \hat{C}(\text{OPT}|\bar{p}) - C(\text{OPT}|\bar{p}) \right| = \left| \sum_{i=1}^{n} \hat{l}_i p_i - \sum_{i=1}^{n} l_i p_i \right|$$

$$\leq \left| \sum_{i=1}^{n} \hat{l}_i p_i - \sum_{i=1}^{n} \hat{l}_i \hat{p}_i \right| + \left| \sum_{i=1}^{n} l_i \hat{p}_i - \sum_{i=1}^{n} l_i p_i \right| .$$

Therefore, for $\delta > 0$ and $0 < \alpha < 1$, it is sufficient to look for the minimal value of m satisfying

$$\text{Prob}\left(\left| \sum_{i=1}^{n} l_i \hat{p}_i - \sum_{i=1}^{n} l_i p_i \right| > \delta \right) < \frac{\alpha}{2} \tag{14}$$

and a similar relation with $\hat{l}_i$ replacing l_i. Now, since $C_i^{(m)} \sim \text{Bin}(m, p_i)$, we can compute the moments of the estimates $\hat{p}$. Using the Chebyshev inequality we have

$$m \leq 2 \frac{\sum_{i=1}^{n} l_i^2 p_i (1 - p_i) - \sum_{i=1}^{n} \sum_{j \neq i} p_i p_j l_i l_j}{\delta^2 \alpha} \leq \frac{2}{\delta^2 \alpha} \sum_{i=1}^{n} l_i^2 p_i (1 - p_i), \tag{15}$$

where the last inequality amounts to neglecting the covariances between the counters. From Lemma 5 we have $l_i \leq 1 - \log p_i / \log \varphi$, hence

$$m \leq \frac{2}{\delta^2 \alpha} \sum_{i=1}^{n} (1 - \log p_i / \log \varphi)^2 p_i (1 - p_i). \tag{16}$$

Each of the terms in the sum is at most $max_{p_i \in (0,1)}(1 - \log p_i / \log \varphi)^2 p_i (1 - p_i) = 1.33371...$ (at $p_i \approx 0.071$), yielding the bound in (11). Observe, that the distribution-free bound obtained for inequality (14) also holds for the inequality

$$\text{Prob}\left(\left|\sum_{i=1}^{n} \hat{l}_i p_i - \sum_{i=1}^{n} \hat{l}_i \hat{p}_i\right| > \delta\right) < \frac{\alpha}{2} . \tag{17}$$

∎

4.2 The Counter Scheme and Weight Balanced Trees

As the computation of the optimal tree requires $\Theta(n^2)$ steps, we are interested in a more efficient construction of *nearly* optimal BSTs. A suitable candidate appears to be the *weight balanced tree*, which is constructed as follows:

Weight Balancing Rule [9]: Choose the root so as to equalize the weight of the left and right subtree as much as possible, then proceed similarly on the subtrees. In fact, it is shown in [7], that a weight balanced tree is constructible in time and space complexity of $\Theta(n)$.

Bayer shows in [3], that for a given RPV $\bar{p}$, the average access cost to the weight balanced tree, denoted by $C(\text{WB}|\bar{p})$ satisfies

$$C(\text{WB}|\bar{p}) - C(\text{OPT}|\bar{p}) \leq lg_2 H + lg_2 e + 1 , \tag{18}$$

where $H = \sum p_i \lg p_i^{-1}$. Since $H \in [0, \log n]$ (the high value is obtained when the RPV is of the uniform distribution), this does look acceptable. We consider below the usage of a scheme, which keeps the tree weight-balanced, i.e. rather than keeping counters and reorganizing the tree only once – at the stopping point – as done by the MOUCS, the tree is modified after each access, so as to keep it weight-balanced by the counters. Denote by $\hat{C}(\text{WB}|\bar{p})$ the cost of the balanced tree constructed by the estimate $\hat{p}$ for the rpv $\bar{p}$.

Theorem 6. *For any $\delta > 0$ and $0 < \alpha < 1$, and for any unknown RPV $\bar{p}$,*

$$Prob\left(\left|\hat{C}(\text{WB}|\bar{p}) - C(\text{WB}|\bar{p})\right| > \delta\right) < \alpha, \tag{19}$$

after a sequence of m_0 accesses to the tree, where

$$m_0 = \frac{2.6675n}{\delta^2 \alpha} . \tag{20}$$

We use in the proof

Lemma 7. *[17] For any set of weights $p_1, \ldots, p_n$ corresponding to the keys $K_1, \ldots, K_n$, such that $S = \sum_{i=1}^{n} p_i \leq 1$, if L_i is the level of K_i in the weight balanced tree (where the level of the root is 0), then*

$$p_i \leq \varphi^{1-L_i} , \tag{21}$$

with φ as defined in Lemma 5.

Proof of Theorem 6: Using the proof of Theorem 3, with $\hat{C}(\text{OPT}|\bar{p})$ and $C(\text{OPT}|\bar{p})$ replaced by $\hat{C}(\text{WB}|\bar{p})$ and $C(\text{WB}|\bar{p})$ respectively, and $\hat{l}_i, l_i$ denoting the levels of K_i in the weight balanced trees for $\hat{p}, \bar{p}$ respectively, m_0 satisfies equation (15). The bound is obtained by using Lemma 7. ■

Corollary 8. *For any $\delta > 0, \alpha < 1$, if the tree is kept weight balanced by the counters till the point m_0 given in (20), then the total number of rotations is bounded by $m_0 \cdot (1 + \lg m_0 / \lg \varphi)$.*

5 Discussion

We have studied reorganization rules for a BST, where accesses to the tree are generated independently by a fixed unknown distribution. We showed that when the distribution is static for sufficiently long durations, the MOUCS:

(i) provides an on-going organization of the tree which improves the expected access cost and requires a low number of rotations,

(ii) yields a search tree with access cost which is arbitrarily close to that of the optimal tree, using statistics accumulated from a reference sequence with length which is linear in the number of elements.

A rule of interest, which reorders the tree as well, while updating the counters, is based on the near-optimal weight balanced tree: During the reference sequence the tree is kept weight-balanced as estimated by the counters. By the monotonic nature of the difference between the estimates and the access probabilities, this rule provides in each stage a closer approximation to the weight-balanced tree constructed for the *known* access probabilities.

In fact, we have shown that for any distribution, the cost of the estimated weight balanced tree approaches the cost of the weight balanced tree of the *unknown* RPV within a number of accesses that is linear in n.

The relative efficiency of the MO rule compared to the scheme which keeps the tree weight balanced by the counters is still open. For long access sequences, we would expect the MO to be inferior with respect to the total average cost of the sequence, but it will retain its advantage of low reorganization cost (the expected number of rotations is $O(n)$).

Consider a quasi-dynamic multiphase environment, where the RPV may change between phases, with K denoting the average length of phases (during a phase the system is static, as in the original model), then we propose the following scheme:

For any reference sequence of length higher than K,

1. Keep counters along the sequence and mark their values every K accesses. The tree is reorganized so as to keep it weight-balanced by the counters. Each phase is a sequence of K requests.

2. After every phase (starting at phase 2), test, whether the counters accumulated in the last two phases where generated by identical distribution. Upon indication that the distribution was changed, reset the counters.

We currently investigate this model and the efficiency of the above scheme. Another issue is the cost of computing (which includes the construction of) the optimal tree. The best known algorithm uses $O(n^2)$ steps and has the same space complexity [14]. The question, whether this algorithm is improvable is still open.

Acknowledgements

We would like to thank Shai Ben-David and Mark Wegman for helpful comments on this paper.

References

1. B. Allen, I. Munro, "Self-Organizing Search Trees", *JACM* 25, #4, 526–535 (1978).
2. M.J. Atallah, S.R. Kosaraju, L.L. Larmore, G.L. Miller, S-II Teng, "Constructing Trees in Parallel", In *Proc. of the 2nd IEEE Symposium on Parallel Algorithms and Architectures*, (1989).
3. P.J. Bayer, "Improved Bounds on the Costs of Optimal and Balanced Search Trees", Tech. Memo. 69, Proj. MAC M.I.T. Cambridge MA 1975.
4. J. Bitner, "Heuristics that Dynamically Organize Data Structures", *SIAM J. Comput.*, 8,1, pp. 82-110, 1979.
5. A. Boneh, M.Hofri, "The Coupon-Collector Problem Revisited." Purdue University, Department of Computer Science, CSD-TR-952, February 1990.
6. W. Feller, *An Introduction to Probability Theory and its Applications* John Willey, New York, 1968.
7. M. L. Fredman, "Two Applications of a Probabilistic Search Technique: Sorting X+Y and Building Balanced Search Trees", *7th ACM Symp. on Theory of Computing*, Albuquerque 1975.
8. I. Galperin, R. Rivest, "Scapegoat Trees", In *Proc. of the 4th ACM-SIAM Symposium on Discrete Algorithms*, Austin, TX, January 25-27, 1993.
9. R. Guttler, K. Mehlhorn, W. Schneider, "Binary Search Trees: Average and Worst Case Behavior", *Jour. of Information Processing and Cybernetics*, 16, 41–61, (1980).
10. M. Hofri, H. Shachnai, "Self-Organizing Lists and Independent References – a Statistical Synergy", *Jour. of Alg.*, 12, 533-555, (1991).
11. M. Hofri, H. Shachnai, "On the Optimality of Counter Scheme for Dynamic Linear Lists", *Inf. Process. Lett.*, 37, 175–179, (1991).
12. T.C. Hu, K. C. Tan, "Least Upper Bound on the Cost of Optimum Binary Search Trees", *Acta Informatica*, 1, 307-310, (1972).
13. D.G. Kirkpatrick, T.M. Przytycka, "Parallel Construction of Binary Trees with Almost Optimal Weighted Path Length", In *Proc. of the 2nd IEEE Symposium on Parallel Algorithms and Architectures*, (1990).
14. D.E. Knuth, "Optimum Binary Search Trees", *Acta Informatica* 1, 14–25,1971.
15. D.E. Knuth, *The Art of Computer Programming, Vol 3: Sorting and Searching* Addison-Wesley, Reading MA 1973.
16. T. W. Lai, D. Wood, "Adaptive Heuristics for Binary Search Trees and Constant Linkage Cost", In *Proc. of the 2nd ACM-SIAM Symposium on Discrete Algorithms*, pp. 72-77, San Francisco, CA, January 28-30, 1991.

17. K. Mehlhorn, "Nearly Optimal Binary Search Trees", *Acta Informatica* **5** 287–295, (1975).
18. K. Mehlhorn, A. Tsakalidis, "Data Structures". In J. van Leeuwen, editor, *Algorithms and Complexity*, Vol A, chapter 6, pp. 301-341, Elsevier, 1990.
19. D.D. Sleator, R.E. Tarjan, "Self-Adjusting Binary search Trees", *JACM* **32**, #3, 652–686 (1985).
20. D.D. Sleator, R.E. Tarjan, "Amortized Efficiency of List Update and Paging Rules", *Commun. ACM* 28,2, pp. 202–208, (1985).

Appendix

Claim 1 *For a given initial* BST, *let* $T_B(I)$ *be the* BST *resulting from processing the reference string* I *with the reorganization rule* B. *Then*

$$T_{MTR}(I) = T_{MO}(I^R),$$

where the string I^R *is the reverse of* I.

Proof: Consider an arbitrary pair of records, R_i and R_j. We look at the sufficient and necessary conditions for R_i to be an ancestor of R_j in $T_{MTR}(I)$ and $T_{MO}(I)$ and call them $C_{MTR}(i,j)$ and $C_{MO}(i,j)$ respectively. The structure of the BST is determined (uniquely) once we specify the ancestor-offspring relation for all pairs of nodes in the tree. Therefore our claim will be established if we show that a string I satisfies $C_{MO}(i,j)$ for all i and j iff $C_{MTR}(i,j)$ holds in I^R for all pairs. The concept of *interval set* is useful for the discussion below. Such a set comprises two records and all other records with keys that lie between them. Let $K_i < K_j$, then the corresponding set is denoted by $IS(i,j)$. Throughout the discussion below we assume w.n.l.g. that $K_i < K_j$ so that the first one of the pair of sets $IS(i,j)$ and $IS(j,i)$ is non-empty. We avoid sticky notation by assuming all records were referenced at least once. For sufficiently long reference strings this holds with an arbitrarily high probability; on the other hand, we should mention that when the *rpv* is far from uniform we expect to obtain informative reference counts long before the above assumption is satisfied. Whatever the case may be, all the claims here hold also for strings that cover only part of the set of records.

The proof devolves from properties of the rotation operation. Refer to Figure 1. We shall say that "the rotated node" is the one that gets to a higher (=lower numbered) level. The second node taking part in the rotation is "the lowered node". The salient properties are:

(a) When a node is rotated it continues to be an ancestor to all of its previous offspring, and *becomes* so to the lowered node and its other subtree.
    The effect of a sequence of rotations of a single node is cumulative.
(b) A lowered node *loses* as offspring the rotated node and its left subtree when the rotation is to the right (or the right one, when the rotation is to the left).

Lemma 9. :

(i) $C_{MO}(i,j)$ is: R_i is the first node to be referenced in $IS(i,j)$.

(ii) $C_{MTR}(i,j)$ is: R_i is the last node to be referenced in $IS(i,j)$.

Proof: The proof of (i) is immediate if we consider the subtree in the initial tree that contains $IS(i,j)$. References (and the consequent rotations) of records outside of this subtree do not change its structure, but may change its level only. References to records in it which are outside of $IS(i,j)$, before R_i is used, will make them ancestors of the entire $IS(i,j)$. Once R_i is referenced (and rotated as high as necessary) it will be ancestor to all other nodes in $IS(i,j)$, and since it will not be lowered again this relation will be maintained indefinitely. Hence the sufficiency.

For the necessity: If some $R_k \in IS(i,j)$, $k \neq i,j$ is referenced before R_i, it will put R_i and R_j in its two separate subtrees, again indefinitely. And lastly, if R_j is the first to be referenced in $IS(i,j)$ it will be the ancestor of R_i.

Part (ii) is due to the fact that a referenced node is rotated all the way to the root. For sufficiency: at the last reference to R_i it reaches the root, and all the rest of $IS(i,j)$ is in its right subtree. Subsequent references to records with lower keys (which are in the left subtree of R_i) will leave it as ancestor of all $IS(i+1,j)$. References to records with keys higher than K_j, will get during their sequence of rotations to have $IS(i+1,j)$ in their left subtrees, and will allow R_i to retain its ancestry with respect to this set (property (b) above). The necessity is similar to the previous case. A subsequent reference to an intermediate key in $IS(i,j)$ will place R_i and R_j in two disjoint subtrees. ∎

The claim is now obvious. ∎

We remark that similar considerations also allow us to determine conditions under which R_i ends up as the *immediate* parent of R_j: in $T_{MO}(I)$ it is required that R_i and R_j were the first two records from $IS(i,j)$ to be referenced, in that order, and the same state will be found in $T_{MTR}(I)$ when R_j and R_i were the last two records referenced, in that order, from $IS(i,j)$.

Time-Message Trade-Offs for
the Weak Unison Problem
(EXTENDED ABSTRACT)

Amos Israeli[1]* Evangelos Kranakis[2]** Danny Krizanc[2]*** and Nicola Santoro[2]†

[1] Technion, Department of Electrical Engineering, Haifa, 32000, Israel
[2] Carleton University, School of Computer Science, Ottawa, ON, K1S 5B6, Canada

Abstract. A set of anonymous processors is interconnected forming a complete synchronous network with sense of direction. Weak unison is the problem where all processors want to enter the same state (in our case "wakeup" state) in the absence of a global start-up signal. As measure of complexity of the protocols considered we use the "bits" times "lag" measure, i.e. the total number of (wakeup) messages transmitted throughout the execution of the protocol times the number of steps which are sufficient in order for all the processors to wakeup. We study trade-offs in the complexity of such algorithms under several conditions on the behavior of the processors (oblivious, non-oblivious, balanced, etc) and provide tight upper and lower bounds on the $time \times \#messages$ measure.
1980 Mathematics Subject Classification: 68Q99
CR Categories: C.2.1
Key Words and Phrases: Anonymous network, Balanced, Chordal rings, t-step protocol, Non-oblivious, Oblivious, Time-message complexity, Unbalanced, Unison, Wakeup protocol, Weak unison.

1 Introduction

An important problem in the current distributed computing literature is the study of the performance of networks when the processors may awaken spontaneously. This paper is concerned with achieving weak unison for all the processors of a synchronous complete network with a sense of direction [12] in the absence of a global start-up signal. The processors have no distinct identities, but the links of the complete network are labeled. Any set of these processors may wake up at any time during the execution of the algorithm. By executing identical protocols and transmitting messages via the network links these "awakened" processors must wake up the entire network. The set of initial processors is arbitrary and may range in size from a single processor to the set of all processors. However, the important point is that the wakeup protocol sought should be such that

---

* amos@ee.technion.ac.il . Partially supported by NWO through NFI Project ALADDIN under contract number NF 62-376.
** kranakis@scs.carleton.ca Research supported in part by NSERC grant.
*** krizanc@scs.carleton.ca . Research supported in part by NSERC grant.
† santoro@scs.carleton.ca . Research supported in part by NSERC grant.

- regardless of the set of processors which are awakened by themselves, eventually all the processors in the network should wake up, and
- the number of messages times the number of steps required for the completion of the algorithm should be optimal.

A weak-unison protocol will specify what the action of each processor will be based on the previous history and the given data at that processor. This means that during the execution of the protocol a set of processors is specified to whom an arbitrary processor, say p, (which already has received wakeup message) should send wakeup messages. Assume that a set of processors is initially awakened. These processors will initiate wakeup messages and will each transmit them to a specified set of processors; in turn the recipient processors retransmit wakeup messages to a new set of processors, and so on until eventually all processors in the network wakeup.

In measuring the quality of the resulting protocol we see an interesting interplay between time required to wake up all the processors in the network and the total number of messages transmitted throughout the execution of the protocol. Thus, a protocol that is time- and message-efficient when many processors initiate wakeup messages may fail to be efficient when there is a single initiator. For example, consider the following protocol in an oriented N-node ring: an initiator sends wakeup message to its left and dies; if a processor is awakened by another processor then it sends wakeup message to its left and dies. It is easy to see that if there is a single initiator the time required to wakeup the whole network is N and the total number of messages is N. However, if all processors are initiators then it takes one time unit for the algorithm to execute and the total number of messages is N. It appears that a fair measure of complexity of this protocol is the time required for all processors to wakeup times the number of bits transmitted throughout the execution of the protocol. Thus the complexity for the former schedule is $\Theta(N^2)$ and for the latter is $\Theta(N)$. In general, we are interested in protocols that have the optimal overall complexity performance regardless of the schedule of initiators.

1.1 Definitions and notations

The set of processors is denoted by $\{0, 1, \ldots, N-1\}$. The network is synchronous and the processors anonymous. The links are labeled, i.e. for any pair $\{x, y\}$ of nodes we associate a label $\ell(\{x, y\})$ such that for any x, if $y \neq y'$ then $\ell(\{x, y\}) \neq \ell(\{x, y'\})$. The orientation we will assume in this paper is the following: for $x, y < N$, the label of the edge $\{x, y\}$ is the integer $y - x \mod N$. This orientation is called in the literature [12] *sense of direction*.

The processors are divided into initiators (those awakened by the adversary) and non-initiators. The former are initiated by themselves, while the latter are awakened.

The number of steps of a wakeup protocol on a given set of initiators is the difference between the time that all processors are awakened and the time the first processor wakes up. The number of messages transmitted in this protocol for a given set of initiators is the number of messages transmitted during this time interval. A protocol is called t-step if this difference never exceeds t regardless of the set of initiators.

An important property of a (correct) wakeup protocol is that all processors must eventually wake up regardless of the schedule (i.e. the configuration of processors which initiate wakeup messages). The nature of a protocol may be such that during its execution a processor may receive, for example, more than one message. In this case the action of the processor involved may or may not depend on its current state. We call the protocol oblivious if the action of each processor does not depend on its current state, but rather it is a function of a predetermined protocol. If for a given protocol, the size of the set to whom each processor transmits messages (during the execution of the protocol) is independent of the processor then the protocol is called balanced.

We are interested in protocols P which minimize the *time* $\times$ *#messages* complexity measure. More precisely, the problem we investigate is the following: "Determine the *time* $\times$ *#messages* complexity for various kinds of wakeup protocols, e.g. oblivious, non-oblivious, balanced, etc."

1.2 Results of the paper

The following table summarizes our bounds for protocols with arbitrary number of steps.

Type of Protocol	Complexity
Arbitrary	$\Omega(N \log \log N)$
Balanced	$\Omega(N \log N)$
Oblivious	$\Omega(N \log^2 N)$
Oblivious	$O(N \log^2 N)$

Clearly, lower bounds valid for arbitrary (respectively, balanced) protocols are also valid for balanced (respectively, oblivious) protocols. Thus our bounds are tight for oblivious protocols (Theorems 10, 9 and Corollary 11) with arbitrary number of steps. Moreover for such protocols the optimal wakeup algorithms are obtained on the hyper-ring architecture (see Example 1).

We also make a detailed analysis of the complexity of t-step protocols and prove the following bounds.

Type of Protocol	# of Steps	Complexity
Arbitrary	t	$\Omega(tN^{2^t/(2^t-1)})$
Balanced	t	$\Omega(tN^{(t+1)/t})$
Oblivious	$t = dN^{1/d}/m$	$O(mdtN)$

The first column describes the type of protocol, the second column gives the number of steps needed by the protocol P, and the last column gives the corresponding *time* $\times$ *#messages*, complexity (henceforth refered to as complexity). As before we see that our bounds are tight for balanced protocols when the number of steps is a constant $t = O(1)$ independent of N. The denominator m in the number of steps of the oblivious case may be any integer $\leq N^{1/d}$, moreover it is easy to see that $O(mdtN) = O(d^2 N^{(d+1)/d})$.

1.3 Related work

Even and Rajsbaum [5, 6] have studied the problem of initialization of computation in (synnchronous as well as asynchronous) distributed networks in the absence of a global start-up signal. They consider the unison problem and study the number of beats it takes for all the processors to be in "unison" (as if they all started the computation at the same time) from the time some processor wakes up. They have shown that a synchronous network of N processors can reach unison within $2N$ beats of the clock [5]. They also show how to achieve unison in $2N$ bits when the network is asynchronous by use of Awerbuch's synchronizer, provided that all local clocks have the same rate [6] (see also [7]). Gouda and Herman [9] present a solution to the unison problem for synchronous systems which is also stabilizing, i.e. the system is guaranteed to reach unison starting from any state. Arora, Dolev and Gouda [1] give a stabilizing solution which for an N-node system uses N registers of $2 \log N$ bits and is guaranteed to converge within N^2 triggers. Covreur, Francez and Gouda [3] give bounded as well as unbounded solutions to unison for asynchronous systems.

Attiya, Snir and Warmuth [2] consider the "processor synchronization problem" on anonymous, synchronous rings in order to reduce input collection and orientation algorithms to the case where all processors start simultaneously and solve this problem in $O(N \log N)$ messages. Fischer, Moran, Rudich and Taubenfeld [8] consider implementations of a related problem, the wakeup problem on a shared register. They give upper and lower bounds on the number of values of the shared register under various assumptions characterizing the resilience of the protocol.

2 Lower Bounds for Arbitrary Protocols

First we consider arbitrary t-step protocols. Usually it is more difficult to prove lower bounds on such protocols because the transmission set of a processor may depend on the input (e.g. the number of wakeup messages received by the processor). Before discussing the general lower bound we present a simpler result for 2-step protocols that better illustrates our proof technique.

2.1 2-step protocols

We will prove an $N^{4/3}$ lower bound for arbitrary 2-step protocols. Throughout $K + x$ denotes the set $\{y + x \bmod N : y \in K\}$, where $K \subseteq \{0, 1, \ldots, N - 1\}$ and $x \in \{0, 1, \ldots, N - 1\}$. First we prove the following lemma.

Lemma 1. *Assume that $K \subseteq \{0, 1, \ldots, N - 1\}$ is a set of size k. There exists a set I of size $\geq \lfloor N/k^2 \rfloor$ such that the sets $K + x$, for $x \in I$, are pairwise disjoint.*

PROOF (OUTLINE) We construct the set I by induction. Assume we have constructed the first s elements $x_0 = 0, x_1, x_2, \ldots, x_{s-1}$ such that the disjointness condition is satisfied. We show how to find the s-th element x_s. We prove that there exists an $x \le sk^2 + 1$ such that $\forall i < s(K + x \cap K + x_i = \emptyset)$. Assume on the contrary that for all $x \le sk^2 + 1$,

$$\exists i < s(K + x \cap K + x_i \ne \emptyset).$$

Then for all $x \le sk^2 + 1$ there exists an $i < s$ and $k_i, k_i' \in K$ such that $x = k_i - k_i' + x_i$. Clearly, there are $\le k^2$ possible differences of elements of K. Hence, the number of elements represented by these last equations is $\le sk^2$, which is a contradiction. It follows that there exists an $x \le sk^2 + 1$ (call this x, x_s) such that

$$\forall i < s(K + x_s \cap K + x_i = \emptyset),$$

as desired. ∎

A set I of processors satisfying the disjointness condition for a set K is called a set of **independent initiators** for K.

Now we are ready to prove an $N^{4/3}$ lower bound on arbitrary 2-step protocols. We have the following theorem.

Theorem 2. $\Omega(N^{4/3})$ *is a lower bound on the complexity of any 2-step wakeup protocol.*

PROOF (OUTLINE) Let us consider a 2-step wakeup protocol. Assume that the protocol is such that it wakes up the whole system under any schedule. By anonymity of the network all initiator processors must execute precisely the same instruction. Hence without loss of generality we may assume that processors initiating wakeup messages transmit to a set K of processors of size k. Now we consider several schedules for waking up the network and study their corresponding complexity.

IF ONLY ONE PROCESSOR WAKES UP:

The processor reaches a set K of k other processors. All these k processors are in the same state and must wake up the remaining $N - k - 1$ processors. Assume that each of them transmits to a set K' of size k' (it must be the same set for all processors since all processors are in the same state). Since all processors must wakeup following this schedule we obtain

$$kk' \ge N - k - 1. \tag{1}$$

IF ALL PROCESSORS WAKE UP:

The total number of messages transmitted will be

$$\ge Nk. \tag{2}$$

IF $\lfloor N/k^2 \rfloor$ INDEPENDENT INITIATORS WAKE UP:

Using the same argument as before, each (independent) initiator reaches k other processors. This wakes up a total of $\frac{N}{k^2}k = \frac{N}{k}$ processors. Since the initiators are independent all these $\frac{N}{k}$ processors are in the same state and will therefore each transmit k' wakeup messages. It follows from (1) that the total number of messages transmitted is

$$\frac{N}{k^2}kk' \geq \frac{N(N-k-1)}{k^2}. \tag{3}$$

The maximum of the quantities in (2), (3) represents a lower bound on the number of messages, namely

$$\max\left\{Nk, \frac{N(N-k-1)}{k^2}\right\} \geq N^{4/3}.$$

This proves the desired lower bound. ∎

2.2 t-step protocols

Next we give a result on the more general case of multistep protocols. This result will require a generaliztion of Lemma 1 regarding the number of independent initiators in an arbitrary wakeup protocol. For $K_1, K_2, \ldots, K_{t-1} \subseteq \{0, 1, \ldots, N-1\}$ define $K = K_1 + K_2 + \cdots + K_{t-1}$ to be the set $\{y_1 + y_2 + \cdots + y_{t-1} : y_i \in K_i,$ for $i = 1, \ldots, t-1\}$. As before we can prove the following result.

Lemma 3. *Assume that $K_1, K_2, \ldots, K_{t-1} \subseteq \{0, 1, \ldots, N-1\}$ are sets of corresponding sizes $k_1, k_2, \ldots, k_{t-1}$. Then there exists a set $I \subseteq \{0, 1, \ldots, N-1\}$ of size $\geq \lfloor N/(k_1 \cdots k_{t-1})^2 \rfloor$ such that the sets $K + x$, for $x \in I$, are pairwise disjoint, where $K = K_1 + K_2 + \cdots + K_{t-1}$.*

PROOF (OUTLINE) Consider the sum set $K_1 + K_2 + \cdots + K_{t-1}$ which has size at most $k_1 k_2 \cdots k_d$ and apply Lemma 1. ∎

This lemma has important implications for the complexity of arbitrary t-step wakeup protocols. We can prove the following theorem.

Theorem 4. $\Omega(tN^{2^t/(2^t-1)})$ *is a lower bound on the complexity of any t-step wakeup protocol.*

PROOF (OUTLINE) Consider the case where only one processor is initiator of a wakeup message. Recall that the network is synchronous and the processors are anonymous. It follows that for each $i = 1, \ldots, t$ there is a set $K_i \subseteq \{0, 1, \ldots, N-1\}$ such that processors awakened at step $i - 1$ transmit wakeup messages to all their neighbors labeled with labels in the set K_i. Let k_i be the size of the set K_i. However the initiator processor must wake up the whole network. This implies that

$$k_1 k_2 \cdots k_t \geq N. \tag{4}$$

Let $i = 1, \ldots, t$ be fixed and I_i be a set of independent initiators "for the first i steps of the protocol" of size $\frac{N}{(k_1 \cdots k_{i-1})^2}$. Such a set exists by Lemma 3 (applied to the sets $K_1, \ldots, K_{i-1}$). If this set of initiators transmits messages following the protocol we obtain the lower bound

$$\frac{N}{(k_1 \cdots k_{i-1})^2}k_1 \cdots k_{i-1}k_i = \frac{N}{k_1 \cdots k_{i-1}}k_i. \tag{5}$$

We claim that inequalities (4) and (5) imply a lower bound $N^{2^t/(2^t-1)}$ on the number of messages. To prove this we argue as follows. If for some $i = 1, 2, \ldots, t$,

$$\frac{k_i}{k_1 \cdots k_{i-1}} \geq N^{1/(2^t-1)}$$

then the claim is proved. Assume on the contrary that for all $i = 1, 2, \ldots, t$ we have that

$$\frac{k_i}{k_1 \cdots k_{i-1}} < N^{1/(2^t-1)}.$$

Using induction we show that $k_i < N^{2^{i-1}/(2^t-1)}$. Indeed,

$$k_i < N^{1/(2^t-1)} k_1 \cdots k_{i-1}$$
$$< N^{1/(2^t-1)} N^{1/(2^t-1)+2/(2^t-1)+\cdots+2^{i-2}/(2^t-1)}$$
$$= N^{2^{i-1}/(2^t-1)}.$$

However this easily contradicts inequality (4). Indeed,

$$k_1 \cdots k_t < N^{(2^0+2^1+\cdots+2^{t-1})/(2^{t-1}-1)} = N^{2^t/(2^{t-1}-1)} < N.$$

The previous proof rests on the fact that for all $i = 1, \ldots, t$ we have that $\frac{N}{(k_1 \cdots k_{i-1})^2} \geq 1$. If this is not true then we need to make some trivial adjustements to the proof. Let i be minimal such that $(k_1 \cdots k_{i-1})^2 > N$. It follows that $(k_1 \cdots k_j)^2 < N$, for $j = 1, \ldots, i-2$. For $j \leq i-2$ consider a set of $\frac{N}{(k_1 \cdots k_j)^2}$ independent initiators. As in the previous proof, the theorem would be proved if for some $j \leq i-2$, $\frac{k_{j+1}}{k_1 \cdots k_j} \geq N^{1/(2^t-1)}$. Hence, without loss of generality we may assume that for all $j \leq i-2$, $\frac{k_{j+1}}{k_1 \cdots k_j} < N^{1/(2^t-1)}$. It follows as before that $k_1 \cdots k_{i-1} < N^{2^{i-1}/(2^t-1)}$, $k_{j+1} < N^{1/(2^t-1)} k_1 \cdots k_j \leq N^{2^j/(2^t-1)}$. This implies that $k_1 \cdots k_{i-1} \leq N^{(2^{i-1}-1)/(2^t-1)} < N^{1/2}$, which contradicts $(k_1 \cdots k_{i-1})^2 > N$. ∎

As a corollary we obtain the following result.

Theorem 5. *The complexity of any wakeup protocol is* $\Omega(N \log \log N)$. ∎

3 Lower Bounds for Balanced Protocols

In this section we give lower bounds which are valid only for balanced protocols. These are protocols for which each processor broadcasts wakeup messages to a set of fixed size. First we consider the case of balanced protocols with fixed number of steps. Note that since oblivious protocols are balanced the lower bound of Theorem 6 is also valid for oblivious protocols. By taking advantage of the the limitations imposed on the problem by the topology we can prove the following lower bound on the complexity of arbitrary balanced protocols.

Theorem 6. *The complexity of every t-step, balanced wakeup protocol is* $\Omega(t N^{(t+1)/t})$.

PROOF (OUTLINE) Consider an arbitrary (non-oblivious) balanced t-step protocol such that in each iteration of the algorithm processors wake up exactly k other processors. If only one processor wakes up then the whole network must also wake up. Hence

$$N \leq k + k^2 + \cdots + k^t \leq 2k^t.$$

It follows that $k \geq \frac{1}{2^{1/t}} N^{1/t}$. If all processors wakeup then a lower bound on the number of wakeup messages is $\Omega(Nk) = \Omega(N^{(t+1)/t})$. Since the time required is t the required complexity is $\Omega(tN^{(t+1)/t})$. This proves the theorem. ∎

As a corollary we obtain the following result.

Theorem 7. $\Omega(N \log N)$ *is a lower bound on the complexity of any balanced, wakeup protocol.* ∎

4 Lower Bounds for Oblivious Protocols

Here we derive a tight lower bound for the special class of oblivious protocols (of arbitrary number of steps). Many of the results on wakeup protocols can be described very naturally within the framework of chordal rings. For this reason next we give some useful notation and definitions on chordal rings.

4.1 Chordal rings

Let Z_N be the set of integers modulo N. Let S be an arbitrary subset of Z_N. The circulant graph $Z_N[S]$ has the nodes $0, 1, \ldots, N-1$ and an edge between the nodes x, y if and only if $x - y \bmod N \in S$. The name "circulant" arises from the fact that the adjacency matrix of $Z_N[S]$ is a circulant matrix.

A related class of graphs are the so-called chordal rings. These are the rings R_N on N nodes, say, such that the set of vertices is $\{0, 1, \ldots, N-1\}$ and there exists a set S such that two nodes x, y are adjacent if and only if $x - y \bmod N \in S$. We denote such a chordal ring by $R_N[S]$. We observe that the well-known ring R_N itself is the chordal ring $R_N[\emptyset]$ and that in general $R_N[S] = Z_N[S \cup \{1\}]$ (this is because the ring structure of R_N assumes the generator 1). Here are two examples of chordal rings [4, 10].

Example 1. For $N \leq k^n$ the chordal ring $R_N[k, k^2, \ldots, k^{n-1}]$ has diameter $\leq k \log_k N$ (use the fact that every $x < N$ can be represented in the basis k) and degree $\log_k N$. If $k = 2$ we call this network the hyper-ring.

Example 2. For $N < n!$ the chordal ring $R_N[2!, 3!, \ldots, (n-1)!]$ has diameter $O(n^2)$ (use the fact that every $x < N$ can be represented in the mixed basis $1!, 2!, \ldots, (n-1)!$ as $x = x_1 + x_2 2! + \cdots + x_{n-1}(n-1)!$, with $0 \leq x_i \leq i$, for $i \geq 1$) and degree $n \leq \log N / \log \log N$.

4.2 Lower bound

Our result is based on the following lemma which gives an $\Omega(\log^2 N)$ lower bound on the product of the diameter times the degree of arbitrary chordal rings. More precisely we have the following result.

Lemma 8. *If the chordal ring* $R_N[k_1, k_2, \ldots, k_{d-1}, k_d]$ *has diameter* δ *then*

$$d \cdot \delta = \Omega\left(\log^2 N\right). \tag{6}$$

PROOF (OUTLINE) Let $x < N$ and suppose that s is the distance of x from 0 in the given chordal ring. A minimal length path connecting x to 0 consists of a_i edges each labeled with k_i, $i = 0, 1, \ldots, d$, where $s = a_0 + a_1 + \cdots + a_d$. It follows that every vertex $x < N$ of the chordal ring corresponds to a partition $i = a_0 + a_1 + \cdots + a_d$ of an integer $i \leq \delta$ into $\leq d + 1$ parts; hence the number N of vertices of the chordal ring cannot exceed the number of partitions of an integer $i \leq \delta$ into $t + 1$ parts. This latter number is majorized by

$$\sum_{i=1}^{\delta} \binom{i+d}{d}.$$

To see this, notice that the mapping

$$(a_0, a_1, a_2, \ldots, a_d) \rightarrow 1^{a_0} 0 1^{a_1} 0 1^{a_2} 0 \cdots 0 1^{a_d}$$

is a $1 - 1$ correspondence between "partitions of i into d parts" and "d-element subsets of a set with $i + d$ elements". Using the well-known identity [11]

$$\sum_{r=0}^{m} \binom{m+r}{r} = \binom{n+m+1}{m}.$$

it follows that

$$N \leq \binom{\delta + d + 1}{\delta}. \tag{7}$$

However the right-hand side of (7) can be majorized using the inequality

$$\binom{x+y}{x} \leq 4^{\sqrt{xy}}. \tag{8}$$

This is is proved by induction on x, for all $y \leq x$. Indeed, it is trivially true for $x = 0$. Assume it is true for x. To prove it for $x + 1 \leq y$ we note that by the induction hypothesis

$$\binom{y+x+1}{x+1} = \frac{y+x+1}{x+1} \binom{y+x}{x} \leq \frac{y+x+1}{x+1} \cdot 4^{\sqrt{xy}}.$$

The right-hand side of this last inequality is easily shown to be $\leq 4^{\sqrt{(x+1)y}}$. Indeed, since $2\sqrt{y/(x+1)} \leq 4^{\sqrt{(x+1)y} - \sqrt{xy}}$ it is enough to show that

$$1 + \frac{y}{x+1} = \frac{y+x+1}{x+1} \leq 2^{\sqrt{y/(x+1)}}.$$

This is equivalent to showing that $1 + t \leq 2^{\sqrt{t}}$, for $0 < t \leq 1$, which in turn is equivalent to proving that $(1 + t)^{\sqrt{t}} \leq 2^t$. But this is trivial since $0 < t \leq 1$. Hence inequality (8) is proved.

It is now clear that inequalities (7) and (8) imply $(\delta + 1)d = \Omega(\lfloor \log N \rfloor^2)$, which in turn implies the lower bound stated in inequality (6). This proves the lemma. ■

Theorem 9. $\Omega(N \log^2 N)$ *is a lower bound on the complexity of any oblivious, wakeup protocol.*

PROOF (OUTLINE) This follows easily from Theorem 8. Since the protocol is oblivious and balanced every processor transmits a fixed number of messages in each iteration of the wakeup protocol, say d. The graph resulting from such a protocol is the chordal ring $R_N[K]$, where K is a set of size d. The time required for the wakeup message to reach all the processors is at least the diameter δ of the chordal ring $R_N[K]$. Eventually all N processors are awakened. Since the protocol is oblivious every processor that receives a wakeup message must transmit to all its d neighbors. Hence the number of messages transmitted during the execution of the protocol is Nd. It follows that the complexity is at least $Nd\delta = \Omega(N \log^2 N)$. ■

5 Upper Bounds

In this section we give wakeup algorithms and prove the upper bounds discussed in subsection 1.2. The protocols we consider here are oblivious. The main theorem is the following.

Theorem 10.

1. *For any d, if $N = k^d$, for some k, then there is an oblivious d-step wakeup protocol whose complexity is $O(d^2 N^{(d+1)/d})$.*
2. *More generally, for any d and any $m \leq N^{1/d}$, if $N = k^d$, for some k, then there is an oblivious $dN^{1/d}/m$-step wakeup protocol whose complexity is $O(mdtN) = O(d^2 N^{(d+1)/d})$.*

PROOF (OUTLINE) To prove the theorem we observe that we can view the d-dimensional mesh as a chordal ring $R_N[K]$, for some set K of links. For example, the 2-dimensional mesh is the chordal ring $R_N[\sqrt{N}]$, while the d-dimensional mesh is the chordal ring $R_N[N^{1/d}, N^{2/d}, \ldots, N^{(d-1)/d}]$. This indicates that wakeup algorithms can be implemented as follows.

First we consider the case of a d-step protocol. For each processor $p = (p_1, p_2, \ldots, p_d)$ let

$$K_p^i = \{(p_1, \ldots, p_{i-1}, x_i, p_{i+1}, \ldots, p_d) : x_i < N^{1/d}\}.$$

If we define $K^i = \{(0, \ldots, 0, x_i, 0, \ldots, 0) : 0 \leq x_i < N^{1/d}\}$ then we see easily that $K_p^i = p + K^i$. Let $K_p = K_p^1 \cup \cdots \cup K_p^d$, and $K = K^1 \cup \cdots \cup K^d$. The protocol is such that processor p transmits wakeup messages to all processors in the set $p + K$. Formally the d-step protocol is as follows.

d-step Wakeup Algorithm
Algorithm for processor p:

1. If p is an initiator then it sends a wakeup message to all its neighbors in the set $p + K$ and dies.
2. If p receives a wakeup message from another processor and is not dead then it sends wakeup messages to all processors in the set $p + K$ and dies.

The size of each broadcast is $dN^{1/d}$. By definition of the protocol a broadcast from processor $p = (p_1, \ldots, p_d)$ will reach all processors of the form $p' = (p_1, \ldots, p_{i-1}, p_i', p_{i+1}, \ldots, p_d)$, where $0 \leq p_i' < N, i = 1, 2, \ldots, d$. Therefore it is clear that every processor will be reached after d steps. The complexity is easily seen to be as in the statement of the theorem.

The $t = dN^{1/d}/m$-step protocol is exactly as before. Each processor p transmits to the set $p + K$, where $K = K_1 \cup \cdots \cup K_d$ and $K_i = \{(0, \ldots, 0, x_i, 0, \ldots, 0) : 0 \leq x_i < m\}$. Details are left to the reader. ∎

As an immediate corollary we obtain the special case of the hyper-ring. This is the chordal ring $R_N[2, 2^2, \ldots, 2^{n-1}]$ described in Example 1 for $N = 2^n$.

Corollary 11. *There is an oblivious $\log N$-step wakeup protocol (implemented on the hyper-ring) whose complexity is $O(N \log^2 N)$.* ∎

We observe that in view of the lower bound for oblivious protocols the result of Corollary 11 is optimal. In addition, Theorem 6 shows that the result of Theorem 10 is also optimal for balanced wakeup protocols with a constant number of steps.

6   Conclusion

We have considered the problem of constructing efficient wakeup protocols on an anonymous, synchronous, complete network. We have constructed oblivious protocols with arbitrary number of steps and shown their optimality, i.e. complexity $\Theta(N \log^2 N)$. For balanced protocols we have given an $\Omega(N \log N)$ lower bound, while for arbitrary protocols an $\Omega(N \log \log N)$ lower bound. This leaves a $\log N$ (respectively, $\log^2 N/\log \log N$) gap for balanced (respectively, arbitrary) protocols from the optimal value for oblivious protocols.

Another interesting question concerns the gap between the $\Omega(N^{4/3})$ lower bound on the complexity of arbitrary 2-step wakeup protocols and the $O(N^{3/2})$ upper bound for oblivious 2-step protocols. A similar question applies to the corresponding lower bound $\Omega(tN^{2^t/(2^t-1)})$ for arbitrary t-step protocols and the upper bound $O(t^2 N^{(t+1)/t})$ for oblivious t-step protocols.

Acknowledgements

Many thanks to J. D. Dixon for suggesting inequality (8) and B. Mans and R. B. Tan for useful conversations.

References

1. A. Arora, S. Dolev and M. Gouda, "Maintaining Digital Clocks in Step", *Proc. of the 5th International Workshop on Distributed Algorithms*, 1991, *Parallel Processing Letters*, Vol. 1, No. 1, pp. 11-18, 1991.
2. H. Attiya and M. Snir and M. Warmuth, "Computing on an Anonymous Ring", Journal of the ACM, 35 (4), 1988. (Short version has appeared in proceedings of the 4th Annual ACM Symposium on Principles of Distributed Computation, 1985, 845 - 875.)
3. J.-M. Couvreur, N. Francez and M. Gouda, "Asynchronous Unison", preprint.
4. P. J. Davis, "Circulant Matrices", John Wiley and Sons, 1979.
5. S. Even and S. Rajsbaum, "Unison in Distributed Networks", in "Sequences, Combinatorics, Compression, Security and Transmission", in *Sequences, Combinatorics, Compression, Security and Transmission*, R.M. Capocelli (ed.), Springer-Verlag, 1990, pp. 479-487.)
6. Unison, Canon and Sluggish Clocks in Networks Controlled by a Synchronizer. To appear in *Mathematical Systems Theory*.
7. S. Even and S. Rajsbaum, "The Use of Synchronizer Yields Maximum Computation Rate in Distributed Networks", STOC 1990, pages 95 - 105.
8. M. J. Fischer, S. Moran, S. Rudich and G. Taubenfeld, "The Wakeup Problem", STOC 1990, pages 106 - 116.
9. M. Gouda and T. Herman, "Stabilizing Unison", Information Processing Letters, Vol. 35, No. 4, pages 171 - 175, 1990.
10. Ki Hang Kim, "Boolean Matrix Theory and Applications", Marcel Dekker Inc., New York, 1982.
11. D. Knuth, "The Art Computer Programming: Fundamental Algorithms", Addison Wesley, 1973.
12. N. Santoro, "Sense of Direction, Topological Awareness and Communication Complexity", ACM SIGACT News, Number 16, pages 50 - 56, 1984.

On Set Equality-Testing

Tak Wah Lam   Ka Hing Lee

Department of Computer Science
University of Hong Kong
Pokfulam Road, Hong Kong
Email address: {twlam, khlee}@csd.hku.hk

Abstract

This paper is concerned with data structures for representing an arbitrary number of sets of keys such that we can update the sets dynamically, query the membership of a set, and test whether two sets are equal. Such data structures have been studied intensively by a number of researchers [2, 6, 4, 5, 7]. Prior to our work, the best scheme supports set equality-testing in constant time and the operations Insert, Delete, and Member each in $O(\log^2 n)$ worst-case time. The solution in this paper improves the $O(\log^2 n)$ time bound to $O(\log n \log^* n)$,[†] while maintaining the constant time complexity of set equality-testing.

1 Introduction

This paper is concerned with a classical problem on representing sets in such a way that set equality-testing can be performed efficiently. More precisely, we want to devise data structures which allow any sequence of the following operations to be performed efficiently.

- Create()—create a new set and return a unique set identifier.

- Insert(i, x)—insert the key x into the set with identifier i.

- Delete(i, x)—delete the key x from the set with identifier i.

- Member(i, x)—return true if the key x is in the set with identifier i, otherwise false.

- Equal(i, j)—return true if the sets with identifiers i and j contain the same set of keys and false otherwise.

[†]$\log^* n = \min\{i \geq 0 \mid \log^{(i)} n \leq 1\}$, where $\log^{(0)} n = n$; for $i > 0$, $\log^{(i)} n = \log(\log^{(i-1)} n)$ if $\log^{(i-1)} n > 0$.

As with previous work on this problem, we assume keys are chosen from an ordered universe and we measure the performance of the data structures in terms of a function of the number of operations performed so far, which is denoted by n below.

Note that if we ignore set equality-testing, the problem is essentially a dictionary problem. That is, we can simply represent each set by a balanced binary search tree (say, an AVL tree) and every operation except equality-testing can be performed in $O(\log n)$ worst-case time. However, such simple solution no longer works if we have to consider equality-testing. More sophisticated data structures are required.

The problem addressed in this paper also has a practical motivation. Modern programming languages such as SETL [3] support sets as primitive data objects and allow testing whether two sets are equal.

Related Previous work and summary of our result: The first nontrivial solution to the problem was given by Sassa and Goto [4] in the 70s. Their solution can perform equality-testing in $O(1)$ time, but it requires $O(n)$ time for the operations Insert, Delete, or Member. A few years later, several researchers succeeded in making use of randomization to improve the linear time complexity. Wegman and Carter [6] devised a randomized solution based on hashing. Every operation can be done in $O(1)$ time, but equality-testing may fail with small probability. Pugh and Tietelbaum [2] showed another randomized scheme which never errs and which supports equality-testing in $O(1)$ time and other operations in $O(\log n)$ *expected* time. Deterministic solutions that can support all operations in polylogarithmic time was not known until the 90s, however. Yellin [7] was the first to devise a deterministic scheme that supports equality-testing in $O(1)$ time and each other operation in $O(\log^2 n)$ worst-case time. Later, Sundar and Tarjan [5] gave another deterministic scheme requiring $O(\log n)$ *amortized* time per operation but $O(\log^2 n)$ time in the worst case.

In this paper, we generalize the work of Yellin [7] to give a deterministic solution which supports equality-testing in $O(1)$ time and other operations in $O(\log n \log^* n)$ worst-case time. In comparison with Yellin's work, our solution is simpler as we avoid using persistent data structures which are fairly complicated to implement. Our improvement of the time complexity is rooted in the introduction of a recursive scheme that can implement *unbounded arrays* efficiently by fixed-size data structures. Conceptually, unbounded arrays are arrays with size not fixed in advance. We show how to manage a number of unbounded arrays such that if the largest index that has been used to access an unbounded array is n, then it always takes $O(\log^* n)$ time to access any entry of that array. We believe that such implementation of unbounded arrays may be useful in other applications.

Organization of this paper: Section 2 gives a brief review of a conceptual data structure called partition trees, which is devised by Yellin [7]. As a warm-up exercise, we restrict our attention to the case in which the maximum number of sets is fixed in advance and we show a new implementation of the partition tree to support set equality-testing in $O(1)$ time, and Member(i, x), Insert(i, x), and Delete(i, x) in $O(\log n)$ time. Section 3 describes how to use fixed-size data structures to implement unbounded arrays efficiently. Section 4 shows how to make use of unbounded arrays to implement the partition tree such that the restriction on the number of sets can be removed with the time complexity of an Insert and Delete increased by a multiplicative factor of $\log^* n$. Finally, we analyze the space required by our data structures.

Preliminaries: Given a memory block B of size s, it is obvious that setting every entry of B to a certain initial value costs $O(s)$ time. However, using a technique from Aho et al.[1], we can avoid such initialization and can still ensure that each entry of B gets the same initial value (say, zero) the first time it is accessed.[‡] Moreover, every entry of B can still be accessed in constant time subsequently. In other words, we can assume that a memory block can be initialized "conceptually" to a certain value in constant time.

In the following discussion, we assume that we never operate on a set that has not been created. Such condition can be checked easily. All the sets created are given integer identifiers sequentially starting from 1. If k sets have been created, they are labeled by $1, 2, \cdots, k$ and denoted by $S_1, S_2, \cdots, S_k$. Note that some of these sets may be empty at some particular moment.

2 Partition Trees

In this section, we first review the definition of partition trees [7]. Then, we consider the special case where the maximum number of sets is fixed in advance and we give a simple implementation that can support the operations Insert, Delete, and Member each in $O(\log n)$ time and set equality-testing in $O(1)$ time.

Consider any sequence of n operations. Suppose that these operations involve k sets $S_1, S_2, \cdots, S_k$ and l distinct keys $x_1, x_2, \cdots, x_l$ for some $k, l \le n$.

[‡]The idea is as follows: We allocate two extra memory blocks Fwd_ptr and $Back\_ptr$ both of size same as B. Intuitively, $Back\_ptr$ is a stack storing all distinct indices that have been used to access B. Let Top be a counter of the number of indices stored in $Back\_ptr$. We maintain the invariant that an entry $B[i]$ has ever been accessed if and only if $Fwd\_ptr[i]$ points to an active region in $Back\_ptr$ (i.e., $Fwd\_ptr[i] \le Top$) and $Back\_ptr[Fwd\_ptr[i]] = i$. Thus, we can always verify in constant time whether the content in $B[i]$ is garbage. The first time an entry $B[i]$ is accessed (i.e., $Fwd\_ptr[i] > Top$ or $Back\_ptr[Fwd\_ptr[i]] \ne i$), we treat $B[i]$ as if it contains the default initial value and we execute the following steps: $Top \leftarrow Top + 1$; $Fwd\_ptr[i] \leftarrow Top$; $Back\_ptr[Top] \leftarrow i$.

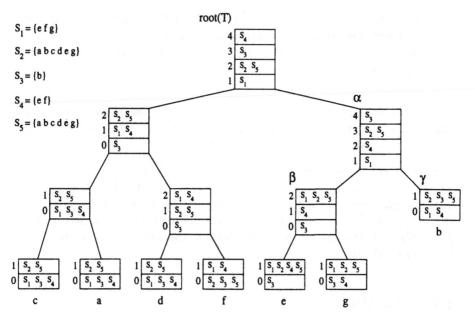

Figure 1: A partition tree for $S = \{S_1, S_2, S_3, S_4, S_5\}$ and $X = \{a, b, c, d, e, f, g\}$. At each node, S is partitioned into several classes each labeled by an integer. Note that $\alpha_1 = \beta_2 \cap \gamma_0$, $\alpha_2 = \beta_1 \cap \gamma_0$, $\alpha_3 = \beta_2 \cap \gamma_1$, $\alpha_4 = \beta_0 \cap \gamma_1$.

Let S and X denote the sets $\{S_1, S_2, \cdots, S_k\}$ and $\{x_1, x_2, \cdots, x_l\}$ respectively.

A partition tree T is an almost complete binary tree with l leaves. Each internal node of T has exactly two children; each leaf corresponds to a distinct key in X and is at depth either $\lceil \log l \rceil$ or $\lceil \log l \rceil + 1$ from the root. For any node α of T, we denote $L(\alpha)$ the set of keys associated with the leaves of the subtree rooted at α and we partition S with respect to α into at most k classes as follows:

> For any sets $S_i, S_j \in S$, they are put in the same class if and only if they contain the same set of keys over $L(\alpha)$ (i.e., $S_i \cap L(\alpha) = S_j \cap L(\alpha)$).

Note that sets having empty intersection with $L(\alpha)$ are all put in the same class. This class, though may be empty, is always labeled by 0. Other classes are labeled by integers chosen from $\{1, 2, \cdots, k\}$. At any node α, a class labeled by an integer r is denoted by α_r.

Figure 1 shows an example of partition trees.

Proposition 1 Let α be any internal node of T. Let $Left(\alpha)$ and $Right(\alpha)$ be α's left and right child respectively. Then, for any $S_i, S_j \in S$, S_i and S_j are in the same class w.r.t. α if and only if S_i and S_j are in the same class w.r.t. both $Left(\alpha)$ and $Right(\alpha)$.

For any internal node α, if the children of α have two classes $Left(\alpha)_p$ and $Right(\alpha)_q$ such that $Left(\alpha)_p \cap Right(\alpha)_q$ is non-empty, there must exist a class α_r at α such that $\alpha_r = Left(\alpha)_p \cap Right(\alpha)_q$.

In the rest of this section, we consider the case in which the number of sets is fixed in advance to be at most a constant c. We show how to implement the partition tree such that the operations Member(i, x), Insert(i, x) and Delete(i, x) can be performed in $O(\log l)$ (i.e., $O(\log n)$) time. In Section 3 and 4, we will study the general case and show that these operations can be done in $O(\log n \log^* n)$ time.

2.1   Data Structures

Suppose the number of sets is fixed in advance to be at most a constant c. Every node α of T is associated with two arrays $Class_\alpha[1..c]$ and $Count_\alpha[1..c]$. Recall that k ($\leq c$) is the number of sets created so far. For any $1 \leq i \leq k$, $Class_\alpha[i] = r$ if S_i belongs to the class α_r. For any non-empty class α_r, $Count_\alpha[r]$ stores the number of sets in α_r. Other entries in these two arrays contain the value zero.

Each internal node α has some additional data structures. $Intersect_\alpha[0..c, 0..c]$ is a table storing the relationship between the classes of α and the classes of its children. For any $0 \leq p, q \leq k$, $Intersect_\alpha[p, q] = r$ if either $p = q = r = 0$, or α_r is a non-empty class at α such that $\alpha_r = Left(\alpha)_p \cap Right(\alpha)_q$; otherwise, $Intersect_\alpha[p, q]$ is said to be undefined. Note that at most $k + 1$ entries in $Intersect_\alpha$ are well defined. Finally, we need some data structures to facilitate the recycling of labels of empty classes: $Largest\_Class_\alpha$ is an integer variable storing the largest class label that have ever been used to label a non-empty class and $Stack_\alpha$ is a stack keeping track of labels of empty classes that are in the range $[1..Largest\_Class_\alpha]$. Whenever we form a new class at α, we try to get a class label from $Stack_\alpha$ before we use the label $Largest\_Class_\alpha + 1$.

Let $Root(T)$ denote the root of the partition tree T. It is obvious that two sets S_i and S_j are equal if and only if at the root of the partition tree T, $Class_{Root(T)}[i] = Class_{Root(T)}[j]$. Thus, equality-testing can be done in constant time. Creating a new set also takes constant time. We simply return $k + 1$ as the next set identifier.

Next, we give the details of updating the partition tree due to an Insert.

2.2 Updating the Partition Tree without creating a new leaf

Consider the operation Insert(i, x) where $1 \leq i \leq k$ and $x \in X$. Let α be the leaf node of T containing the key x. For any node γ of T, if $L(\gamma)$ does not contain x, the partitioning of sets at γ remains the same after Insert(i, x) is processed. In other words, only those nodes on the path from α to the root of T are to be updated. In the following, we show that such nodes can be updated in a bottom-up manner.

Basis: The leaf node α has at most two non-empty classes: α_0 contains all the sets in which x is absent and α_1 contains all other sets. Insert(i, x) causes the set S_i to move from α_0 to α_1. Thus, $Class_\alpha[i]$ should be set to 1 and $Count_\alpha[1]$ is incremented by one.

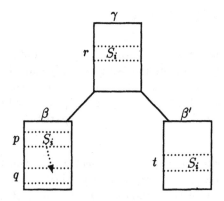

Figure 2: S_i has moved from β_p to β_q.

Inductive Step: Suppose we have updated a node β on the path from α to $Root(T)$. Assume S_i has moved from a class β_p to another class β_q. Let γ and β' be respectively the parent and sibling of β. See figure 2. Without loss of generality, we assume that β is the left child of γ. At this point, let $r = Class_\gamma[i]$ and let $t = Class_{\beta'}[i]$. That is, just before Insert(i, x) is executed, S_i was in the classes γ_r and β'_t. As mentioned earlier, S_i should remain in the class β'_t. Yet, with respect to γ, S_i should move from γ_r to another class γ_s such that $\gamma_s = \beta_q \cap \beta'_t$.

- If β_q and β'_t had some sets in common before Insert(i, x) is executed (i.e., $Intersect_\gamma[q, t]$ is not undefined), we let $s = Intersect_\gamma[q, t]$ and insert S_i into γ_s.

– Otherwise, $Intersect_\gamma[q, t]$ is undefined. Let $s \geq 1$ be a label of an empty class (which can be found in constant time through the data structures $Stack_\gamma$ and $Largest\_Class_\gamma$). S_i is the only set to be put into γ_s.

In either case, the updating of γ due to Insert(i, x) includes setting $Class_\gamma[i]$ and $Intersect_\gamma[q, t]$ to s, and modifying the values of $Count_\gamma[r]$, $Count_\gamma[s]$, $Stack_\gamma$, and $Largest\_Class_\gamma$. Moreover, when $Count_\gamma[r]$ gets down to zero, $Intersect_\gamma[p, t]$ should be marked undefined.

The procedure described above requires constant time to update each node on the path from the leaf α to $Root(T)$. Since the height of T is $\lceil \log l \rceil + 1$, Insert(i, x) costs $O(\log l)$ (i.e., $O(\log n)$) time.

Delete(i, x) can also be processed in a similar fashion in $O(\log n)$ time.

2.3   Creating a New Leaf

Consider the operation Insert(i, x) where $1 \leq i \leq k$ and $x \notin X$. That is, no leaf of T contains the key x. In the following, we only show how to add a leaf into T for x in $O(1)$ time. Insert(i, x) can then be performed in a way as described in the previous section. Unlike Yellin's work [7], we avoid using persistent data structures when we add a new leaf into the partition tree. This makes our solution simpler.

To add a new leaf α for x, we first select a leaf β in the partition tree T. As T is an almost complete binary tree, every leaf is at depth either $\lceil \log l \rceil$ or $\lceil \log l \rceil + 1$. To maintain the balance of T, we select β from leaves at depth $\lceil \log l \rceil$ whenever possible. Let γ be the parent of β. Now we create two new nodes ρ and α such that α is a leaf containing the key x and ρ is an internal node occupying the current position of β. Note that β and α become the left child and the right child of ρ respectively. See Figure 3.

At this point, all the sets in S (including S_i) does not contain x. Adding a leaf for x into T does not change the partitioning of the sets and the content of the data structures in every existing node of T. It remains to build the data structures for ρ and α before we can process Insert(i, x).

With respect to α, all sets should be in class α_0. Thus, we allocate the arrays $Class_\alpha[1..c]$ and $Count_\alpha[1..c]$ with all entries initialized to 0.

At β, sets are partitioned into at most two classes β_0 and β_1; at α, all sets are in the same class α_0. Thus, sets are partitioned at ρ in a way exactly the same as β. That is, ρ has at most two non-empty classes ρ_0 and ρ_1 which are equal to β_0 and β_1 respectively. Moreover, at most two entries in the array

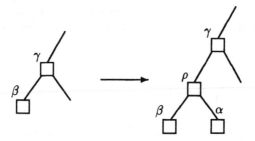

Figure 3: Inserting α and ρ into the partition tree

$Intersect_\rho[1..c, 1..c]$, namely $Intersect_\rho[0, 0]$ and $Intersect_\rho[1, 0]$, are well defined. We allocate the array $Intersect_\rho[1..c, 1..c]$ with all entries initialized to be "undefined", and then assign $Intersect_\rho[0, 0] = 0$, and $Intersect_\rho[1, 0] = 1$ if β_1 is non-empty.

The initialization of the data structures $Largest\_Class_\rho$ and $Stack_\rho$ is also trivial. The only data structure that would cause trouble is $Class_\rho$. $Class_\rho[i]$ should be equal to $Class_\beta[i]$ for all $1 \leq i \leq k$, yet copying $Class_\beta$ to $Class_\rho$ may cost $O(k)$ time. In Section 4, this may become a problem as we have to address the case where k is not bounded by any fixed constant. Fortunately, as to be explained below, there is a way to avoid the brute-force copying and the time required is a constant not depending on k.

Instead of copying $Class_\beta[1..c]$ to $Class_\rho[1..c]$ directly, we keep a pointer W_ρ at ρ pointing to β. W_ρ will be kept unchanged in the future. When the array $Class_\rho$ is allocated, every entry $Class_\rho[i]$ is initialized with a special symbol "useless" to indicate that the actual content of $Class_\rho[i]$ is stored in $Class_\beta[i]$.

In general, whenever we access $Class_\tau[i]$ of an internal node τ, we must first check whether $Class_\tau[i] =$ "useless". If so, let $\omega = W_\tau$ and copy the value stored in $Class_\omega[i]$ to $Class_\tau[i]$. Note that $Class_\tau[i]$ is no longer "useless". On the other hand, for a leaf λ, there is no such pointer W_λ and $Class_\lambda[i]$ never contains the "useless" symbol. It is possible that up to $\lceil \log l \rceil$ internal nodes of T, each of which is an ancestor of λ, have their pointers pointing to λ. When we want to alter the value in $Class_\lambda[i]$, we must first copy the old value to each $Class_\tau[i]$, where τ is an ancestors of λ with $W_\tau = \lambda$ and $Class_\tau[i] =$ "useless".

In summary, adding a leaf into T for a new key costs only $O(1)$ time but at the expense of complicating the access of the array $Class$ at every node of T. Although accessing an entry in the array $Class$ of an internal node can still be done in $O(1)$ time, it requires $O(\log l)$ time to access an entry in $Class_\lambda$ of a leaf λ. Fortunately, according to the algorithms in the Section 2.2, when we actually process an Insert or Delete, we alter only one entry in the array *Class* of one leaf. Thus, the time required to process an Insert or Delete is still $O(\log l)$ (i.e., $O(\log n)$) time.

2.4 Membership Testing

To support the operation Member(i, x), we keep another AVL tree U to store all the keys in X. More precisely, every node in U corresponds to a key $x \in X$ and it stores a pointer to a leaf α of the partition tree T containing x. U is ordered in respect of the keys. With U, any query Member(i, x) can be processed in $O(\log l)$ time as follows: Search U for x. If U does not contain a node for x, report Member$(i, x) = $ false; otherwise, let α be the leaf in T containing x. Report Member$(i, x) = $ false if $Class_\alpha[i] = 0$, or else report true.

To maintain U, every time a leaf is added to T, we also add the corresponding node to U. This can also be done in $O(\log l)$ (i.e., $O(\log n)$) time.

In fact, the algorithms presented in Section 2.2 and 2.3 also rely on U to locate a leaf in T for any given key.

3 Implementation of Unbounded Arrays

In Section 2, we have seen that the size of the data structures at each node of the partition tree is determined by the number of sets. If the maximum number of sets is fixed in advance, all these data structures càn be implemented by a number of fixed-size arrays. Otherwise, we need to use *unbounded* arrays (i.e., arrays without any constraint on the size). Of course, we don't know how to allocate the space for a number of unbounded arrays in advance. In this section, we show how to use fixed-size data structures to implement *unbounded* arrays efficiently. Then, we would be able to implement the partition tree without restricting the number of sets.

To be more specific, an unbounded array $A[1..]$ is an array of which the size is not fixed in advance. Intuitively, A can grow up dynamically to meet the size required. All entries of A should contain the same initial value (say, zero) and every entry of A is to be accessed as if A is a fixed-size array. Note that there is no restriction on the pattern of access to A and A may be a sparse array. For instance, the first one hundred access to A are all restricted to the first ten entries of A, yet the next access can possibly be to $A[1000]$.

At any particular time, we define the *size* of an unbounded array A to be the largest index that has been used to access A so far.

Obviously, one can use an AVL tree to implement an unbounded array A. For every entry $A[i]$ that has been accessed, there is a node in the AVL tree containing the index i and the value in $A[i]$. The tree is ordered in respect of the indices. At any time, if the current size of A is m, there can be at most m nodes in the AVL tree and an access to A requires $O(\log m)$ time in the worst case. This solution is simple but inefficient. In the following, we show two other schemes to implement an unbounded array A by some fixed-size data structures. The first scheme is simpler; it ensures that if the current size of A is m, the implementation of A requires $O(m)$ space and each access of A can be processed in $O(\log \log m)$ time. The second scheme is a recursive version of the first one and improves the access time to $O(\log^* m)$ time.

Scheme I: To implement an unbounded array A, we use an AVL tree I and a list of memory blocks $\mathcal{B} = (B_{i_1}, B_{i_2}, B_{i_3}, \cdots)$, where $1 \leq i_1 < i_2 < i_3 < \cdots$. Each block B_i is of size 2^{i-1}; it corresponds to the $(2^{i-1})th$, $(2^{i-1}+1)th$, ..., $(2^i - 1)th$ entries of A. No two blocks in $\mathcal{B}$ are of the same size. $\mathcal{B}$ is empty initially. The block B_i is allocated only when we first access an entry $A[j]$ with $2^{i-1} \leq j < 2^i$. Again, we assume all entries in B_i can be initialized to a particular value in constant time. Let us look at an example. Suppose we have only accessed $A[3]$, $A[18]$, and $A[23]$. $\mathcal{B}$ should consist of blocks B_2 and B_5. The tree I keeps track of the actual locations of the arrays in $\mathcal{B}$. For each array B_i, there is a node in I storing the index i and a pointer to the first entry of B_i. I is ordered with respect to the index of each block.

To access the jth entry of A, we first compute $i = \lfloor \log j \rfloor + 1$.§ That is, $2^{i-1} \leq j < 2^i$. We search I to check whether there is a block B_i. If it exists, we access the entry $B_i[j - 2^{i-1} + 1]$ as $A[j]$. Otherwise, we allocate the block B_i which is of size 2^{i-1}, and insert a node into I for B_i.

Suppose m is the current size of A. The largest block in $\mathcal{B}$ is of size at most m and the number of blocks in $\mathcal{B}$ is bounded by $\lfloor \log m \rfloor + 1$. The height of I is $O(\log \log m)$. It takes $O(\log \log m)$ time to search I or insert a node into I. Therefore, any access to an entry of A can be processed in $O(\log \log m)$ time. Moreover, the space used by I and $\mathcal{B}$ is always dominated by the size of the largest block, which is $O(m)$.

Scheme II: Instead of using an AVL tree I, we use another unbounded array I' to keep track of the blocks in $\mathcal{B}$. Entries in I' are supposed to be

---

§Obviously, we cannot compute $\lfloor \log j \rfloor$ in constant time for any arbitrary j. Nevertheless, when we make use of unbounded arrays to implement the partition tree, such j is always bounded by the number of sets k created. We can build a logarithmic table incrementally as the sets are created. That is, when the set S_k is created, we fill in the value of $\lfloor \log k \rfloor$ into the table.

undefined initially. When the block B_i is created for some $i \geq 1$, we store in $I'[i]$ a pointer to the first location of B_i. Whenever we want to access the jth entry of A, we compute $i = \lfloor \log j \rfloor + 1$ and retrieve the pointer to B_i from $I'[i]$. If $I'[i]$ is undefined, we allocate the block B_i and update $I'[i]$ accordingly. Thus, the time to access an entry of A is bounded by that of I' plus some constant.

Let m be the current size of A. There are at most $\lfloor \log m \rfloor + 1$ blocks in B and the size of I' is at most $\lfloor \log m \rfloor + 1$. If we implement I' by an AVL tree and another list of memory blocks, then an access to I' and hence an access to A can be processed in $O(\log \log \log m)$ time.

In fact, the idea above can be applied recursively to obtain a scheme with $O(\log^* m)$ access time.

Definition: Let $f(x)$ be an integer function such that $f(x) = \lfloor \log x \rfloor + 1$. For any $i \geq 0$, define $f^{(i)}(x) = x$ if $i = 0$, otherwise $f(f^{(i-1)}(x))$.

An unbounded array A is to be implemented as a multi-level structure. The number of levels depends on the current size m of A. More specifically, there are p levels where p is the smallest integer such that $f^{(p)}(m) \leq 2$. Note that $p = O(\log^* m)$. Each level t consists of a list of memory blocks $\mathcal{B}^t = (B_{i_1}^t, B_{i_2}^t, B_{i_3}^t, \cdots)$, where $1 \leq i_1 < i_2 < i_3 < \cdots$ and each block B_i^t is of size 2^{i-1}. The toppest level $\mathcal{B}^p$ has exactly two blocks B_1^p and B_2^p. $\mathcal{B}^1$ is the actual data structure storing the entries of A. For all $t > 1$, $\mathcal{B}^t$ stores pointers to keep track of the blocks in $\mathcal{B}^{t-1}$.

To access the jth entry of A, we first compute $l_1 = f^{(1)}(j)$, $l_2 = f^{(2)}(j)$, $\cdots$, $l_p = f^{(p)}(j)$. If $f^{(p)}(j) \leq 2$, we try to follow the pointers stored in $B_{i_p}^p[l_{p-1} - 2^{l_p} + 1]$, $B_{i_{p-1}}^{p-1}[l_{p-2} - 2^{l_{p-1}} + 1]$, $\cdots$, $B_{i_2}^2[l_1 - 2^{l_2} + 1]$ to access the block $B_{i_1}^1$ at level 1. $A[j]$ is stored in $B_{i_1}^1[j - 2^{l_1} + 1]$. In the course of tracing the pointers, if we encounter an undefined entry at a particular level t, we allocate a sequence of blocks one for each level $t' < t$.

If $f^{(p)}(j) > 2$, we need to add extra levels to the current data structures. Let p' be the smallest integer such that $f^{(p')}(j) \leq 2$. On top of $\mathcal{B}^p$, $p' - p$ levels $\mathcal{B}^{p+1}, \mathcal{B}^{p+2}, \cdots, \mathcal{B}^{p'}$ are added. For $p \leq t < p'$, $\mathcal{B}^t$ consists of three blocks B_1^t, B_2^t, and $B_{f^t(j)}^t$. At the toppest level, $\mathcal{B}^{p'}$ has two blocks $B_1^{p'}$ and $B_2^{p'}$ only. Now we can proceed as if we encounter an undefined entry at level p.

Again, we assume that $f^{(i)}(j)$ can be computed in constant time by looking up some table. The time complexity of accessing an entry of A depends only on the number of levels, which is $O(\log^* m)$. The space required by the whole scheme is dominated by the space used by the lowest level (i.e., level 1), which is $O(m)$.

4 No constraint on the number of sets

The solution is still based on the partition tree. As the maximum number of sets is not fixed in advance, we can no longer allocate fixed-size arrays to implement the data structures *Class*, *Count*, and *Intersect* at each node of the partition tree. Instead, we use unbounded arrays.

Suppose we have already processed a sequence of n operations. If k sets have been created, the size of the unbounded arrays required for *Class* and *Count* is at most k and an access to either array requires $O(\log^* k)$ time. Recall that *Intersect* is a two-dimensional table. We implement it by an unbounded array of unbounded arrays. Each entry in the first-level unbounded array is a pointer to an unbounded array of the second level. In either level, the size of any unbounded array is at most k. Thus, an access to an entry of *Intersect* still requires $O(\log^* k)$ time.

Since the time to access an entry of the data structures *Class*, *Count*, and *Intersect* at any node is increased by a multiplicative factor of $O(\log^* k)$ $(= O(\log^* n))$, the operations Insert(i, x) or Delete(i, x) can be performed in $O(\log n \log^* n)$ time. Note that Member(i, x) only takes $O(\log n + \log^* n)$ $(= O(\log n))$ time.

To have $O(1)$ time set equality-testing, we must be able to access any entry of $Class_{Root(T)}$ in $O(1)$ time. We use a fixed-size array R (instead of an unbounded array) to represent $Class_{Root(T)}$. Initially, R is an array with two entries. The first time we create a class with label greater than the size of R, we allocate another array R' of size doubled. We will not copy the content of R to R' immediately as this may take more than constant time. Instead, whenever we create another class later, we spend constant time to copy one entry from R to R'. Note that we need both R and R' to serve an access to $Class_{Root(T)}$. By the time when we request a class label greater than the size of R', all entries of R should have been copied to R'. Now, R is no longer useful and we can replace R by R'. At any time, if k sets have been created, the space required by the current R and R' plus their predecessors is $O(k)$. Note that the trick of representing $Class_{Root(T)}$ cannot be extended to the data structures of all the nodes of the partition tree.

5 Space Complexity

Consider any sequence of n operations. Recall that l and k denote respectively the number of distinct keys and sets involved. Note that there are $O(l)$ nodes in the partition tree T.

At each node α, the space required by all data structures excluding $Intersect_\alpha$ is $O(k)$. Summing over all the nodes of T, these data structures use $O(kl)$ (i.e., $O(n^2)$) space. For the data structure $Intersect$, each Insert operation causes the allocation of at most $O(k)$ space for each node on the path from the leaf containing x to $Root(T)$. Note that the height of T is $O(\log l)$. The total space allocated due to an Insert operation is $O(k \log l)$. Therefore, a sequence of n operations uses $O(nk \log l)$ (i.e., $O(n^2 \log n)$) space.

References

[1] A.V. Aho, J.E. Hopcroft, and J.D. Ullman, *The Design and Analysis of Computer Algorithms*, Ex 2.12, 71-71.

[2] W. Pugh and T. Teitelbaum, Incremental Computation via Function Caching, *Proceedings of the Sixteenth ACM Symposium on Principles of Programming Languages*, 1989, 315-328.

[3] J.T. Schwartz, On Programming: An Interim Report on the SETL Project, Installments I and II, *CIMS New York University*, 1974.

[4] M. Sassa and E. Goto, A Hashing Method for Set Operations, *Information Processing Letters*, 5, 1976, 265-279.

[5] R. Sundar and R.E. Tarjan, Unique Binary Search Tree Representations and Equality-testing of Sets and Sequences, *Proceedings of the Twenty-second Annual ACM Symposium on Theory of Computing*, 1990, 18-25.

[6] M.N. Wegman and J.L. Carter, New Hash Functions and Their Use in Authentication and Set Equality, *Journal of Computer and System Sciences*, 22, 1981, 265-279.

[7] D.M. Yellin, Representing Sets with Constant Time Equality Testing, *Journal of Algorithms*, 13, 1992, 353-373; a preliminary version also appeared in the *Proceedings of the First Annual ACM-SIAM Symposium on Discrete Algorithms*, 1990, 64-73.

On the Complexity of Some Reachability Problems

Angelo Monti and Alessandro Roncato

Dipartimento di Informatica, Universitá di Pisa
Corso Italia 40, 56125 Pisa, Italy
{montico,roncato}@di.unipi.it

Abstract. In this paper we introduce the notion of rewriting systems for sets, and consider the complexity of the reachability problems for these systems, showing that this problem is PSPACE-complete in the general case and is P-complete for particular rewriting systems. As a consequence, we show that the emptiness and finiteness problems for E0L systems are PSPACE-complete, solving in this way an open problem. Finally, we give completeness results for some decision problems concerning binary systolic tree automata.

1 Introduction

In this paper we introduce rewriting systems for sets, which we call Set Systems. We investigate the complexity of some problems arising from these systems. In particular we show that the reachability problem is PSPACE-complete: given any Set System, determine whether specific elements can be reached by applying the rules of the system. The infinitely reachability problem is also PSPACE-complete: the problem of determining whether these elements can be reached infinitely many times. By putting some restriction on the kind of production that are allowed, we obtain the Propagating Set Systems reachability problem. We can show that such problem is P-complete. Using the above results we close a long standing open problem (see [12, 14, 15, 16]) showing that emptiness problem for E0L system is PSPACE-complete. Moreover we prove that the finiteness problem for E0L is PSPACE-complete too. Finally, we show that emptiness, finiteness and equivalence problems for Binary Systolic Tree Automata are PSPACE-complete.

The rest of this paper is organized into five parts. In Section 2 we define our rewriting systems and prove the PSPACE-completeness of our reachability problems. In Section 3 we prove the P-completeness of the reachability problem for propagating rewriting systems. The above results will be used in Sections 4 and 5 to classify the complexity of the emptiness and finiteness problems for E0L systems and the complexity of emptiness, finiteness and equivalence problems for Binary Systolic Tree Automata. Finally in Section 6 we summarize our results, and their implications on previous work, suggesting new directions for further research.

2 Set Systems

A Set System consists of a finite set P_S of productions of the form $X \to Y$ where X and Y are subsets of a finite set S. A Set System is in Normal Form iff for each $X \to Y, X' \to Y' \in P_S$ if $X = X'$ then $Y = Y'$. Note that for a Set System P_S it is easy to define an equivalent one P_S' where productions are defined $X \to \bigcup_{X \to Y_i \in P_S} Y_i$. So, without loss of generality, in the following we speak of Set Systems, meaning Set System in Normal Form. A set $A \subseteq S$ derives in one step (in P_S) a set $B \subseteq S$, written $A \Rightarrow_{P_S} B$, if and only if $B = \bigcup_{X \to Y \in P_S, X \subseteq A} Y$.

We write $A \overset{i}{\Rightarrow}_{P_S} B, i \geq 1$, when A derives B in i steps and $\overset{*}{\Rightarrow}_{P_S}$ is the transitive and reflexive closure of $\Rightarrow_{P_S}$.

The reachability problem for Set Systems (RP) is

$\text{RP} = \{< S_0, P_S, S^F > | S_0 \overset{*}{\Rightarrow}_{P_S} S', S' \cap S^F \neq \emptyset\}$.

The Infinitely Reachability Problem (IRP) is

$\text{IRP} = \{< S_0, P_S, S^F > | \text{ for each } n \in N \text{ there is an } i > n \text{ such that } S_0 \overset{i}{\Rightarrow}_{P_S} S', S' \cap S^F \neq \emptyset\}$.

A Set system is said to be a 2-Set System iff for each $X \to Y \in P_S$ holds $|X| = 2$ and $|Y| \leq 2$.

R2P and IR2P are the above problems restricted to 2-Set Systems.

Lemma 1. *RP and IRP are in PSPACE*

Proof. Consider the instance $< S_0, P_S, S^F >$ of RP (or IRP). We run the following algorithm:

1. start with $A = S_0$ and $i = 0$;
2. if $A \cap S^F \neq \emptyset$ then answer "YES" and stop;
3. if $i > 2^{|S|}$ then answer "NO" and stop;
4. compute $A = \bigcup_{X \to Y \in P_S, X \subseteq A} Y$, $i = i + 1$ and go to step 2.

Examining $2^{|S|}$ consecutive sets starting from S_0, at least one set appears twice in the sequence of A, hence the sequence is cyclically repeated from this set. If none of the elements of S^F appears in the first $2^{|S|}$ elements of the sequence, it will not appear in the following. Looking at the space necessary to above algorithm we have to store:

- the set A which has at most cardinality $|S|$;
- a temporary copy of the set A (to compute the next value of A);
- The number i which has the value at most $2^{|S|}$ and can be stored using $\log 2^{|S|} = |S|$ space.

Then we need only a polynomial working space. To answer at the question posed by IRP we can use the following algorithm:

1. start with $A = S_0$ and $i = 0$;

2. if $A \cap S^F \neq \emptyset$ and $i > 2^{|S|}$ then answer "YES" and stop;
3. if $i > 2^{|S|+1}$ then answer "NO" and stop;
4. compute $A = \bigcup_{X \to Y \in P_S, X \subseteq A} Y$, $i = i + 1$ and go to step 2.

If one elements of S^F appears between the sets $2^{|S|}$ and $2^{|S|+1}$ of the sequence then it appears in a cyclic part of the sequence, that is it appears infinitely many times. If none of the elements of S^F appears between these sets this means that or it appears only before the $2^{|S|}$-th set, then it appears only finitely many times, or it never appears. The valuation of the necessary space is like the precedent one.

Lemma 2. *R2P and IR2P are PSPACE-Hard.*

Proof. We start showing that every problem in PSPACE is polynomial time reducible to R2P. Let $M = (Q, \Sigma, \Gamma, \delta, q_0, \square, F)$ be a semi-infinite one tape and polynomial space DTM defined as in [11].

For simplicity assume that the function δ is defined for all its arguments. Noting that for a given input w the tape head of M may scan only the first n tape cells with $n = p(|w|)$ and p a polynomial function, we may code an instantaneous description (ID) $x_1 \ldots x_{i-1} q x_i x_{i+1} \ldots x_n$ of M as a set
$C(x_1 \ldots x_{i-1} q x_i x_{i+1} \ldots x_n) =$
$= \{(x_1, 0, 1), \ldots, (x_{i-1}, 0, i-1), (x_i, q, i), (x_{i+1}, 0, i+1), \ldots, (x_n, 0, n)\}$
where $q \in Q$ and $x_j \in \Gamma$ for $1 \leq j \leq n$.
Here we use:

- the element $(x_j, 0, j)$ to denote that the j-th tape cell contains the symbol x_j and the tape head is elsewhere.
- the element (x_j, q, j) to denote that the tape head is on the j-th cell containing the symbol x_j and that the state of finite control is q.

Now we give a polynomial time algorithm that takes as input a string w and produces an instance $< S_0, P_S, S^F >$ for RP such that S_t, $t \geq 0$, is the code of the ID of M working on w at time t.

Let $w = w_1 \ldots w_{|w|}, w_i \in \Sigma, 1 \leq i \leq |w|$. Then $< S_0, P_S, S^F >$ is defined as:

- $S = \{(x, q, i) : x \in \Gamma, q \in Q \cup \{0\}, 1 \leq i \leq n)\}$;
- $S_0 = \{(w_1, q_0, 1)\} \cup \{(w_i, 0, i) : 2 \leq i \leq |w|\} \cup \{(\square, 0, i) : |w| + 1 \leq i \leq n\}$;
- $S^F = \{(x, q, i) : x \in \Gamma, q \in F, 1 \leq i \leq n\}$
- For each $q \in Q$ and $a \in \Gamma$, P_S is defined as follows:
  - if $\delta(q, a) = (p, b, Right)$ then for each $i \in \{1, \ldots, n-1\}$ and $x, y, z \in \Gamma$:
    * $\{(a, q, i), (y, 0, i+1)\} \to \{(b, 0, i), (y, p, i+1)\}$;
    * $\{(z, 0, j), (a, q, i)\} \to \{(z, 0, j)\}$ for each $j \in \{1, \ldots, n\} \setminus \{i, i+1\}$.
  - if $\delta(q, a) = (p, b, Left)$ then for each $i \in \{2, \ldots, n\}$ and $x, y, z \in \Gamma$:
    * $\{(x, 0, i-1), (a, q, i)\} \to \{(b, 0, i), (x, p, i-1)\}$;
    * $\{(z, 0, j), (a, q, i)\} \to \{(z, 0, j)\}$ for each $j \in \{1, \ldots, n\} \setminus \{i-1, i\}$.

It is easy to see that the above reduction is polynomial in $|w|$, the length of input string. We can also prove that such a reduction can be done in LOG-SPACE. First note that the construction of sets S_0 and S^F can be done within LOG-SPACE (we only need to store a counter from 1 to n, and the value n). For the construction of P_S we have to store the numbers i and j of the production we are defining, plus the value of n. All other information is of finite nature.

We now prove that the set S_t, $t \geq 0$, derived by S_0 in P_S in t steps is the code of the ID of M working on w at time t.

The proof is by induction:

- $\underline{t = 0.}$ By definition of Initial ID of a DTM and by our definition of S_0, this latter set is the code of the Initial ID of M.
- $\underline{t \Rightarrow t + 1.}$ By inductive hypothesis we have that S_t is the code of the ID of M at time t. We have to prove that S_{t+1} is the code of next ID.
  It is easy to see that assuming $C(x_1 \ldots x_{i-1} q a x_{i+1} \ldots x_n) = S_t$
  - if $\delta(q, a) = (p, b, Right)$ then $S_t \Rightarrow_{P_S} C(x_1 \ldots x_{i-1} b p x_{i+1} \ldots x_n)$;
  - if $\delta(q, a) = (p, b, Left)$ then $S_t \Rightarrow_{P_S} C(x_1 \ldots p x_{i-1} b x_{i+1} \ldots x_n)$.

Therefore we have proved that, for any t, S_t represents the t-th ID of M. This machine reaches a final ID iff $(x, q, i) \in S_j$ for some $j \in N, i \in \{1, \ldots, n\}, x \in \Gamma$ and $q \in F$. That is, there exists $j \in N$ such that $S_j \cap S^F \neq \emptyset$.

Hence R2P is PSPACE-Hard; now we have to prove that also IR2P is PSPACE-Hard. Starting with the DTM M, it is easy to (effectively) construct a DTM M' such that M reaches a final state iff M' reaches a final state infinitely many times. Then the proof follows likewise as for RP using M' instead of M.

From the Lemma 1 and Lemma 2 we have:

Theorem 3. *RP and IRP are PSPACE-complete problems.*

3   Propagating Systems

A Propagating Set System is such that for every element $s \in S$ we have $\{s\} \rightarrow \{s\} \in P_S$.

PRP ($PIRP$) is the RP (IRP) problem for Propagating Set Systems.

Note that in a Propagating Set System P_S when an element is reached then it is reached infinitely many times. Hence $PRP = PIRP$.

Lemma 4. *PRP is in P*

Proof. In a Propagating Set System $X \overset{*}{\Rightarrow}_{P_S} Y$ implies that $Y = X \cup E, E \subseteq S$. Hence we may use the algorithm for RP with the test "$i > |S|$" instead of "$i > 2^{|S|}$".

Lemma 5. *The problem PRP is P-Hard.*

Proof. We must show that every problem in P is LOGSPACE reducible to PRP. Let $M = (Q, \Sigma, \Gamma, \delta, q_0, \Box, F)$ be a semi-infinite one tape and polynomial time DTM.

Let p the polynomial function bounding the steps of M. Noting that for a given input w the tape head of M may scan only the first n tape cells with $n = p(|w|)$, we may code an instantaneous description (ID) $x_1 \ldots x_{i-1} q x_i x_{i+1} \ldots x_n$ of M at time t as a set

$C(x_1 \ldots x_{i-1} q x_i x_{i+1} \ldots x_n) =$
$= \{(x_1, 0, 1, t), \ldots, (x_{i-1}, 0, i-1, t), (x_i, q, i, t), (x_{i+1}, 0, i+1, t), \ldots, (x_n, 0, n, t)\}$
where $q \in Q$ and $x_j \in \Gamma$ for $1 \leq j \leq n$ $0 \leq t \leq n$

Here we use:

- the element $(x_j, 0, j, t)$ to denote that at time t the j-th tape cell contains the symbol x_j and the tape head is elsewhere.
- the element (x_j, q, j, t) to denote that at time t the tape head is on the j-th cell containing the symbol x_j and that the state of finite control is q.

Now we give a polynomial time algorithm that takes as input a string w and produces an instance $< S_0, P_S, S^F >$ for PRP such that $S_0 \overset{t}{\Rightarrow}_{P_S} S_t, 0 \leq t \leq n$, implies that S_t contains the code of the ID at time t. Namely $S_t = \bigcup_{j=0}^{t} C(ID_j)$ where ID_j is the ID of M working on w at time j.

Let $w = w_1 \ldots w_{|w|}, w_i \in \Sigma, 1 \leq i \leq |w|$. Then $< S_0, P_S, S^F >$ is defined as:

- $S = \{(x, q, i, t) : x \in \Gamma, q \in Q \cup \{0\}, 1 \leq i \leq n, 0 \leq t \leq n)\}$;
- $S_0 = \{(w_1, q_0, 1, 0)\} \cup \{(w_i, 0, i, 0) : 2 \leq i \leq |w|\} \cup \{(\Box, 0, i, 0) : |w| + 1 \leq i \leq n\}$;
- $S^F = \{(x, q, i, t) : x \in \Gamma, q \in F, 1 \leq i \leq n, 0 \leq t \leq n\}$
- For each $s \in S$, $0 \leq t \leq n - 1$, $q \in Q$ and $a \in \Gamma$, P_S is defined as follows:
  - $\{s\} \rightarrow \{s\}$
  - if $\delta(q, a) = (p, b, Right)$ then for each $i \in \{1, \ldots, n - 1\}$ and $x, y, z \in \Gamma$:
    * $\{(a, q, i, t), (y, 0, i+1, t)\} \rightarrow \{(b, 0, i, t+1), (y, p, i+1, t+1)\}$;
    * $\{(z, 0, j, t), (a, q, i, t)\} \rightarrow \{(z, 0, j, t+1)\}$
      for each $j \in \{1, \ldots, n\} \setminus \{i, i+1\}$;
  - if $\delta(q, a) = (p, b, Left)$ then for each $i \in \{2, \ldots, n\}$ and $x, y, z \in \Gamma$:
    * $\{(x, 0, i-1, t), (a, q, i, t)\} \rightarrow \{(b, 0, i, t+1), (x, p, i-1, t+1)\}$;
    * $\{(z, 0, j, t), (a, q, i, t)\} \rightarrow \{(z, 0, j, t+1)\}$
      for each $j \in \{1, \ldots, n\} \setminus \{i-1, i\}$.

It is easy to see that the above reduction can be done in LOG-SPACE. First note that the construction of sets S, S_0 and S^F can be done within LOG-SPACE (we only need to store two counters from 1 to n, and the value n). For the construction of P_S we have to store the number i, j and t for the production we are defining, plus the value of n. All other information is of finite nature. We now prove that for each set S_t, $t \geq 0$, derived by S_0 in P_S in t steps holds: $S_t = \bigcup_{j=0}^{t} C(ID_j)$ where ID_j is the ID of M working on w at time t. the code of the ID of M working on w at time t.

The proof is by induction:

- $t = 0$. By definition of Initial ID of a DTM and by our definition of S_0, this latter set is the code of the Initial ID of M.
- $t \Rightarrow t+1$. By inductive hypothesis we have that: $S_t = \bigcup_{j=0}^{t} C(ID_j)$. We have to prove that $S_{t+1} = S_t \cup C(ID_{t+1})$.

  It is easy to see that assuming $C(x_1 \ldots x_{i-1} q a x_{i+1} \ldots x_n) = ID_t$
  - if $\delta(q,a) = (p,b,Right)$ then $S_t \Rightarrow_{P_S} C(x_1 \ldots x_{i-1} b p x_{i+1} \ldots x_n) \cup S_t$;
  - if $\delta(q,a) = (p,b,Left)$ then $S_t \Rightarrow_{P_S} C(x_1 \ldots p x_{i-1} b x_{i+1} \ldots x_n) \cup S_t$.

Therefore we have proved that, for any t, S_t contains only the first t-th IDs of M. This machine reaches a final ID iff $(x,q,i,t) \in S_j$ for some $j \in N, i, t \in \{1, \ldots, n\}$, $x \in \Gamma$ and $q \in F$. That is, there exists $j \in N$ such that $S_j \cap S^F \neq \emptyset$.

From the above two lemmas we have:

Theorem 6. *PRP is a P-complete problem.*

4 Decision problems for E0L systems

Several authors have studied the complexity of several decision problems for L systems, showing that these problems, are complete for well known complexity classes. See [14] for an overview.

In this section we study the emptiness and finiteness problems for E0L systems. For previous works on these problems see [12], [15] and [16].

We will show that both problems are PSPACE complete. In fact we prove a slightly stronger result, that is they are PSPACE-complete also for the restricted case of EP0L systems.

First we recall the relevant definitions from the theory of L systems and then give the proofs of PSPACE completeness.

An E0L system is a construct $G = (V, T, P, \omega)$ where:

- V is a finite alphabet,
- $T \subseteq V$ is called the target alphabet,
- $\omega \in V^+$ is a word, called *axiom*,
- $P \subset V \times V^*$ is a finite binary relation such that, for every symbol $a \in V$, there exists $\alpha \in V^*$ such that $\langle a, \alpha \rangle \in P$. If for each $\langle a, \alpha \rangle \in P$ holds that $\alpha \neq \epsilon$ then the E0L system is called EP0L system.

Let $x = a_1 \ldots a_k$, $k \geq 1$, $a_1, \ldots, a_k \in V$, and $y \in V^*$. We say that x *derives* y in one step (in G), $x \Rightarrow_G y$, if there exists $\alpha_1, \ldots, \alpha_k \in V^*$ such that $\langle a_i, \alpha_i \rangle \in P$ for $1 \leq i \leq k$ and $y = \alpha_1 \ldots \alpha_k$. We write $x \overset{i}{\Rightarrow}_G y$ when x derives y in i steps. $\overset{*}{\Rightarrow}_G$ is the transitive and reflexive closure of $\Rightarrow$. The language of G is defined as $L(G) = \{x \in T^* | \omega \overset{*}{\Rightarrow}_G x\}$.

Lemma 7. *The emptiness and finiteness problems for EP0L systems are PSPACE-hard.*

Proof. We show that RP is polynomially reducible to the nonemptiness problem for EP0L, and IRP is polynomially reducible to the infiniteness problem for EP0L. The thesis then follows from Lemma 2.

Let $I = < S_0, P_G, S^F >$ be an instance of RP or IRP. Consider the EP0L system $G = (V, T, P, \omega)$ such that:

- $V = S \cup \{r, s\}$, where r and s are two symbols not in S;
- $T = S_0$;
- $P = \{< y, x_1 \ldots x_n > |X \to Y \in P_S, y \in Y, X = \{x_1, \ldots, x_n\}\} \cup \{\langle x, rr \rangle | x \in S\} \cup \{\langle r, rr \rangle\}$

  (the last two sets guarantee that, for every symbol in V, there is a pair in P);
- $\omega = s$.

It is easy to show by induction on t that, for each $q \in S$, we have:

$q \overset{t}{\Rightarrow}_G w \in T^+$ iff $S_0 \overset{t}{\Rightarrow}_{P_S} S', q \in S'$;

therefore we have:

$L(G) \neq \emptyset$

$\quad \Leftrightarrow$ there exists $t \geq 1$ and $w \in T^+$ such that $s \overset{t}{\Rightarrow}_G w$

$\quad \Leftrightarrow$ there exists $t \geq 1$, $w \in T^+$ and $q \in S^F$ such that $s \Rightarrow q \overset{t-1}{\Rightarrow}_G w$

$\quad \Leftrightarrow$ there exists $t \geq 1$ such that $S_0 \overset{t-1}{\Rightarrow}_{P_S} S', q \in S' \cap S^F$

$\quad \Leftrightarrow < S_0, P_S, S^F > \in RP$

$\quad L(G)$ is infinite

$\quad \Leftrightarrow$ for each $t \geq 1$ there exists a $j \geq t$ and $w \in T^+$ such that $s \overset{j}{\Rightarrow}_G w$

$\quad \Leftrightarrow$ for each $t \geq 1$ there exists a $j \geq t$, $q \in S^F$ and $w \in T^+$ such that

$s \Rightarrow_G q \overset{j-1}{\Rightarrow}_G w$

$\quad \Leftrightarrow$ for each $t \geq 1$ there exists a $j \geq t$ such that $S_0 \overset{j-1}{\Rightarrow}_{P_S} S', q \in S' \cap S^F$

$\quad \Leftrightarrow < S_0, P_S, S^F > \in IRP$

To complete the proof is sufficient to note that G can be constructed in polynomial time starting from I and also in LOG-SPACE.

Theorem 8. *The emptiness and the finiteness problems for E0L systems are PSPACE-complete.*

Proof. It is known that these problems are in PSPACE, see [12], so the result follows from the above lemma.

5 Decision problems for systolic tree automata.

Culik et al. [1] have introduced the notion of systolic tree automata as a simple model for VLSI computation. A number of results concerning binary systolic tree automata (BSTA's) were established in [2, 3, 4, 5, 6, 7]. On one hand the class of languages recognized by such devices is large enough to contain all regular languages and, in addition, other quite complicated languages. On the other hand, the class is small enough to possess a number of important decidability properties [2, 6]. In this section we show that the problems of emptiness, finiteness and equivalence for BSTA are all PSPACE-complete.

A BSTA $K = (\Sigma, Q, F, g, h)$ consists of an infinite binary tree without leaves, an input alphabet Σ, a finite set of working states Q with a special state $\sharp$ not in Σ, a set $F \subseteq Q$ of final states, two functions $g : \Sigma \cup \{\sharp\} \to Q$ and $h : Q \times Q \to Q$. Given a input word $w \in \Sigma^*$ of length t, we choose the first level in the tree with $n \geq t$ vertices. The word $w\sharp^{n-t}$ is 'fed' character by character to the level in question and the nodes in the level are labeled with the g- values of the characters. Now the information flows bottom-up and in parallel, and a node is labeled $h(a, b)$ when its left and right sons have been labeled a and b, respectively.

The word w is accepted by K if the root of the tree is labeled with a value from F. The language recognized by K is the set $L(K)$ of all the accepted words.

Lemma 9. *The nonemptiness, infiniteness and equivalence problems for BSTA are in PSPACE.*

Proof. Consider the BSTA $K = (\Sigma, Q, F, g, h)$ and the sequence $S_i = (A_i, B_i, C_i, D_i)$ where:

- $A_0 = B_0 = C_0 = \{g(x)|x \in \Sigma\}$,
- $D_0 = \{g(\sharp)\}$,
- $A_{i+1} = \{h(x, y)|x, y \in A_i\}$,
- $B_{i+1} = \{h(x, y)|x \in A_i, y \in C_i\}$,
- $C_{i+1} = \{h(x, y)|x \in C_i, y \in D_i\} \cup \{h(x, y)|x \in A_i, y \in C_i\}$,
- $D_{i+1} = \{h(x, x)|x \in D_i\}$.

Let $p = |Q|$. From the results in [2] we have:

1. $L(K) \neq \emptyset$ iff there exists an r, $0 \leq r \leq p2^{2p}$, such that $B_r \cap F \neq \emptyset$,
2. $L(K)$ is infinite iff there exists an r, $p2^{2p} \leq r \leq 2p2^{2p}$, such that $B_r \cap F \neq \emptyset$.

Noting that the sequence $S_i = (A_i, B_i, C_i, D_i)$ may be generated storing only two consecutive $S_i's$, and that each of the sets in S_i may be stored as a bit vector of size p, it is easy to see that the conditions contained in points 1 and 2 can be tested in in linear space. Therefore the thesis holds for the nonemptiness and infiniteness problems.

The decidability of the equivalence is proven in [2] by noting that $L(K_1) = L(K_2)$ iff $(\overline{L(K_1)} \cap L(K_2)) \cup (L(K_1) \cap \overline{L(K_2)}) = \emptyset$, and recalling that the class of the languages recognized by BSTA's is (effectively) closed under the Boolean operations.

It is known that BSTA's recognizing $\overline{L(K_i)}$, $L(K_1) \cup L(K_2)$ and $L(K_1) \cap L(K_2)$ may be constructed in polynomial time, starting from K_1 and K_2, [1]. Hence it is easy to see that the procedure:

- construct the automaton B such that $L(B) = (\overline{L(K_1)} \cap L(K_2)) \cup (L(K_1) \cap \overline{L(K_2)})$;

- verify that $L(B) = \emptyset$;

for the equivalence problem requires polynomial space.

Theorem 10. *The emptiness and the finiteness problems for BSTA are PSPACE-complete.*

Proof. We show that problems $R2P$ and $IR2P$ are polynomially reducible to the nonempty and to the infiniteness problems, respectively. Hence the thesis follows from the PSPACE-hardness of $R2P$ and $IR2P$, and from the Lemma 9. Let $< S_0, P_S, S^F >$ be an instance for $R2P$ or $IR2P$. Construct a BSTA $K = (\Sigma, Q, F, g, h)$ such that $Q = S \cup \{\natural\}$, $\Sigma = S_0$, $F = S^F$, g is the identity over $\Sigma \cup \{\natural\}$, and

- $h(p,q) = p', h(q,p) = q'$ when $\{p,q\} \to \{p',q'\} \in P_S$
- $h(p,q) = p', h(q,p) = p'$ when $\{p,q\} \to \{p'\} \in P_S$
- $h(p,q) = \natural$ otherwise.

Clearly the construction of K from $< S_0, P_S, S^F >$ may be done in polynomial time. Moreover the automaton K has the following two properties:

- $L(K)$ does not contain words whose length is not a power of two;
- $S_0 \stackrel{j}{\Rightarrow}_{P_S} S', q \in S'$ iff there exists a word over Σ of length 2^j that labels the root of the tree the state q.

The first property holds because, by definition of h and g with respect to the symbol $\natural$, this symbol, when present at the input level, is propagated bottom-up to label the root. Note also that $\natural$ is present in all strings whose length is not a power of two, and $\natural$ is not in F.

The second property may be proven by an easy induction on j.

Now it is easy to see that $L(K) \neq \emptyset$ iff $< S_0, P_S, S^F > \in R2P$. In fact:

$L(K) \neq \emptyset$

$\Leftrightarrow$ there exists a w, $|w| = 2^j$, such that, when w is given as input to K, it labels the root with a state in F (i.e. in S^F)

$\Leftrightarrow$ there exists a j such that $S_0 \stackrel{j}{\Rightarrow}_{P_S} S', S^F \cap S' \neq \emptyset$.

$\Leftrightarrow < S_0, P_S, S^F > \in R2P$

So we have given a polynomial reduction from $R2P$ to the nonemptines problem for BSTA.

Analogously we may prove that $L(K)$ is infinite iff $< S_0, P_S, S^F > \in IR2P$. In fact:

$L(K)$ is infinite

$\Leftrightarrow$ there exists an infinite sequence of words over Σ, $\{w_i\}_{i \geq 0}$, $|w_i| = 2^{j_i}$, $j_u \neq j_v$, $u \neq v$, and each of these words, when given as input to K labels the root with a state in F (i.e. in S_F)

$\Leftrightarrow$ for each j there exists a j', $j' \geq j$, such that $S_0 \stackrel{j'}{\Rightarrow}_{P_S} S', S' \cap S^F \neq \emptyset$.

$\Leftrightarrow < S_0, P_S, S^F > \in IR2P$

Theorem 11. *The equivalence problem for BSTA is PSPACE-complete.*

Proof. Using the trivial reduction of the emptiness problem to the equivalence problem, and the results in Lemmas 9 and 10.

It is useful remark here that the above PSPACE-hardness results are in a sense stronger than the ones given in the previous section. In fact an alternative way to obtain these last results would be to recall that the class of BSTA languages is a proper subclass of the E0L languages, see [3], and to show that, from a BSTA automaton K, it is possible construct in polynomial time an E0L system generating the language L(K). However we preferred to give the results on E0L systems without passing trough the BSTA model, this choice has permitted to obtain the theorem for EP0L.

6   Concluding remarks

The main result of this work is perhaps the proof of PSPACE-completeness for the emptiness problem for E0L systems. Several authors have worked on this problem. In [12] emptiness was proved to be decidable in polynomial space. In [15, 16] it was shown that this problem is complete for the class of all $NARPAC(\log n)$ languages, (i.e. the class of languages accepted by $RPAC$ automata introduced by Rozenberg in [17]) which are augmented with a LOG-SPACE working tape and a two way input head.
Note that this problem is located between the emptiness problem for ED0L (i.e. E0L systems such that for each $a \in V$ there is exactly one pair in P with a as first component) and the emptiness problem for ET0L (i.e. E0L systems generalized so that instead of a single relation P there is a finite set of them and at each derivation step one of them is nondeterministically chosen). In [12] it was proved that these two problems are respectively NP-complete and PSPACE-complete thereby establishing the relation:

$$NP \subseteq NARPAC(\log n) \subseteq PSPACE.$$

By our result we conclude that $NARPAC(\log n) = PSPACE$.
The question whether it is possible to locate the emptiness problem of E0L systems in polynomial hierarchy was posed in [14] as an open problem. Our result makes this possibility very unlikely.
For a BSTA $K = (\Sigma, Q, F, g, h)$, our results imply that a number of important properties for K, although decidable, are probably computationally intractable (unless $P = PSPACE$). On the other hand, putting particular restrictions on the type of allowed transition functions for K it is possible to prove that all the above problems become decidable in polynomial time.
Finally noting that fast recognition of regular languages by BSTA is of special interest, we pose the open question: Are there particular restrictions on the transition functions for BSTA such that the resulting subfamily of languages has the nice property to contain all the regular languages and still it is possible decide

the emptiness, finiteness and equivalence problems in polynomial time? (Observe that the restriction to be propagating does not allow the recognition of regular languages)

Another interesting question is: generalizing the Set Systems to Multisets Systems, is still the reachability problem decidable?

References

1. K. Culik II, A. Salomaa and D. Wood: Systolic Tree Acceptors. R.A.I.R.O. Inform. Theor. 18, 53-69 (1984)
2. K. Culik II, J. Gruska and A. Salomaa: Systolic Automata for VLSI on balanced tree. Acta Informatica 18, 335-344 (1983)
3. K. Culik II, J. Gruska and A. Salomaa: On a Family of L languages Resulting from Systolic Tree Automata. Theoret. Comp. Sci. 23, 231-242 (1983)
4. E. Fachini, A. Maggiolo Schettini, G. Resta and D. Sangiorgio: Some structural properties of systolic tree automata. Fundamenta Informaticae 12, 571-586 (1989)
5. E. Fachini, A. Maggiolo Schettini, G. Resta and D. Sangiorgio: Nonacceptability criteria and closure properties for the class of languages accepted by binary systolic tree automata. Theoretical Computer Science 83, 249-260 (1991)
6. E. Fachini and A. Monti: Regular sets, context-free languages and Systolic Y-tree Automata. In Proc. of the 4-th Italian conference on Theoretical Computer Science. L'Aquila: World Scientific 1992, pp. 199-213
7. E. Fachini and A. Monti: A Kleene-like characterization for languages accepted by Systolic Tree Automata. J. of Computer and systems sciences *to appear*.
8. J. Gruska: Synthesis, structure and power of systolic computations. Theoretical Computer Science 71, 47-77 (1990)
9. G. T. Herman and G. Rozenberg: Developmental Systems and Languages. North-Holl., Amsterdam 1975
10. O. K. Ibarra and S. M. Kim: A characterization of systolic binary tree automata and applications. Acta Informatica 21, 193-207 (1984)
11. J.E. Hopcroft and J.D. Ullman: Introduction to Automata Theory, Languages and Computation. Addison-Wesley, Reading Mass. 1979.
12. N. Jones and S. Skyum: Complexity of some problems concerning L systems. Math. Systems Theory, 13, 29-43 (1979)
13. K.-J. Lange: L systems and NLOG-reductions. In: G. Rozenberg and A. Salomaa (editors): The Book of L. Berlin: Springer-Verlag 1986, pp.45-91.
14. K.-J. Lange: Complexity Theory and Formal Languages. In: Proc. of the 5-th IMYCS, LNCS 381, Berlin: Springer-Verlag 1989, pp.19-36.
15. K.-J. Lange and M. Schudy: A further link between formal languages and complexity theory. EATCS bull., 33, 33-67 (1979)
16. K.-J. Lange and M. Schudy: The Complexity of the Emptiness Problem for E0L Systems. In G. Rozenberg and A. Salomaa (editors): Lindenmayer Systems. Berlin: Springer-Verlag 1992, pp. 166-175.
17. G. Rozenberg: On a family of acceptors for some classes of developmental languages. Internat. J. Comput. Math. 4, 199-228 (1974)
18. G. Rozenberg and A. Salomaa: The Mathematical Theory of L systems. New York: Academic Press 1980

On self-reducible sets of low information content

(Extended Abstract)

Martin Mundhenk*

Universität Trier, Fachbereich IV - Informatik, D-54286 Trier, Germany

Abstract. Self-reducible sets have a rich internal structure. The information contained in these sets is encoded in some redundant way. Therefore a lot of the information of the set is easily accessible. In this paper it is investigated how this self-reducibility structure of a set can be used to access easily *all* information contained in the set, if its information content is small. It is shown that P can be characterized as class of self-reducible sets which are "almost" in P (i.e. sets in APT'). Self-reducible sets with low instance complexity (i.e. sets in IC[log, poly]) are shown to be in $NP \cap co\text{-}NP$, and sets which disjunctively reduce to sparse sets or which belong to a certain superclass of the Boolean closure of sets which conjunctively reduce to sparse sets are shown to be in P^{NP}, if they are self-reducible in a little more restricted sense of self-reducibility.

1 Introduction

The internal properties of a set have effects on its computational complexity and the usability of the information encoded into the set. For example, a "natural" encoding of the satisfiable Boolean formulas over an alphabet Σ of at least two letters – i.e. the set SAT – is NP-complete[2]. But also one can consider a tally encoding of SAT over a one-letter-alphabet (say *tally*(SAT)), or an encoding which does not contain any redundant information (*compress*(SAT)). Using the properties of SAT that it has exponential density, that SAT is decidable in exponential time, and that SAT contains a lot of redundant information, it follows under the assumption $P \neq NP$, that whether *tally*(SAT) (cf. [Ber]) nor *compress*(SAT) (cf. [BL]) are hard for NP, which means that the information contained in these "extreme" encodings is not usable by an efficient algorithm to decide SAT. These arguments give lower and upper bounds of the information content of SAT.

---

* Work done at Universität Ulm, Abt. Theoretische Informatik. Supported in part by the DAAD through Acciones Integradas 1992, 313-AI-e-es/zk. Author's e-mail address: mundhenk@uni-trier.de

[2] for example take the binary representation of Boolean formulas in a LaTeX text as encoding

The redundancy of SAT can be expressed in terms of *self-reducibility*:[3] a Boolean formula F having variables $x_1, \ldots, x_n$ is satisfiable iff $F|_{[x_1/true]}$ is satisfiable or $F|_{[x_1/false]}$ is satisfiable, where $F|_{[x_1/a]}$ describes the formula obtained from F where the variable x_1 is replaced by the value $a \in \{true, false\}$, and which therefore has variables $x_2, \ldots, x_n$. Since the satisfiability of every formula having no variables can be decided in polynomial time, one can imagine a decision procedure for $F \in$ SAT? which procedes deciding $F|_{[x_1/true]} \in$ SAT? and $F|_{[x_1/false]} \in$ SAT? using "recursive function calls", where the computation of the arguments for each call and the evaluation of the results is polynomially time bounded, and the depth of the recursion is bounded by the number of variables of F. If a set of low information content has such redundancy properties, one can derive certain upper bounds for the complexity of the set. This was used to show that P = NP, if SAT bounded truth-table reduces to a sparse set [OW], or if SAT is contained in the Boolean closure of the sets which conjunctively reduce to sparse sets [AHH+]. For more general types of reducibilities, Karp and Lipton [KL] showed the collaps of the Polynomial Time Hierarchy PH on $\mathrm{NP^{NP}}$, if an NP-complete set Turing reduces to a sparse set. In this paper for example the disjunctive reducibility – whose strength relative to sparse sets lies intermediate between bounded truth-table and Turing reducibility – is considered, and it is shown that if SAT disjunctively reduces to a sparse set, then PH collapses to $\mathrm{P^{NP}}$.

The general form of these results can be seen as: if a set is self-reducible and reduces in a certain way to a sparse set, then its complexity is low. Lozano and Torán [LT] studied the lowness of self-reducible sets in the class of sets which many-one reduce to sparse sets. In this paper upper bounds of the complexity of self-reducible sets in several classes defined by various reducibilities to sparse sets are considered. At first a natural general definition of self-reducibility is given, where the only restriction on the self-reducibility is that the arguments of the "recursive calls" do not leave a polynomially bounded range. We will show that a restriction of this type of self-reducibility covers all of the self-reducibility types used up to now. We give characterizations of the classes which contain polynomial time "approximizable" sets as APT (the class of sets which are "almost" in P [MP]) and P-*close* [Sch2] in terms of reduction classes to sparse sets. It will be shown that all self-reducible sets in APT are in P, and that self-reducible sets of small instance complexity (i.e. sets in IC[log, poly]) are in NP ∩ *co*-NP. For the restricted type of self-reducibility it is shown that self-reducible sets which disjunctively reduce to sparse sets or which are in the Boolean closure of sets which conjunctively reduce to sparse sets are low in Δ_2^p. An overview of the results is given in a figure in Section 5. It shows the relationships between the considered classes, their containments in the Extended Low Hierarchy (defined by [BBS]), and the lowness results in case of self-reducibility as shown in this paper.

---

[3] see Section 2 for a formal definition

2 Preliminaries and notation

In general the notations and definitions from the standard books on structural complexity theory (for example, [BDG, Sch2]) are used.

Let A be a set. $A^{=n}$ ($A^{\leq n}$) denotes the set of all strings in A of length n (up to length n, respectively). The cardinality of A is denoted by $|A|$. A set S is called sparse if for some polynomial p and every n $|A^{\leq n}| \leq p(n)$. We use SPARSE and co-SPARSE to represent the classes of sparse sets and sets having a sparse complement, respectively. A set S is called C-printable if the function which maps 0^n to an encoding of $S^{\leq n}$ can be computed in C. A set A is said to be *low in* C for a complexity class C, if $C^A \subseteq C$ (see [BBS]). $\langle \cdot, \cdot \rangle$ denotes some easy computable and invertible pairing function. $L(M)$ is the set accepted by the Turing machine M, $L(M, A)$ is the set accepted accepted by the oracle Turing machine M using the set A as oracle.

The reducibilities discussed in this paper are standard polynomial time bounded reducibilities as *conjunctive reducibility* $\leq_c^p$, *disjunctive reducibility* $\leq_d^p$ (see [BDG]), and the Hausdorff reducibility $\leq_{hd}^p$ by Wagner [Wag]. Note that $A \leq_c^p B$ if and only if $\overline{A} \leq_d^p \overline{B}$. For any reducibility $\leq_r^\alpha$ and any class C of sets let $R_r^\alpha(C) = \{A \mid A \leq_r^\alpha B \text{ for some } B \in C\}$.

The classes APT (almost polynomial time [MP]) and P-*close* [Sch2] contain sets which only on few instances differ from sets in P and can intuitively be seen as "approximizable" in P. The following defined class APT' is easily seen to be a generalization of APT. ($A \triangle B$ denotes the symmetric difference between sets A and B.)

Definition 1 *A set A is in* P-close, *if $A = L(M) \triangle S$ for a deterministic polynomially time bounded Turing machine M and a sparse set S. If furthermore $S \subseteq S'$ for a sparse $S' \in$ P, then A is in* APT'.

The notion of instance complexity and the class IC[log,poly] of sets of strings with low instance complexity were introduced by Ko, Orponen, Schöning, and Watanabe (see [KOSW]). Remark that IC[log, poly] $= R_c^p(\text{SPARSE}) \cap R_d^p(\text{co-SPARSE})$ (cf. [AHH+, BLS]). An overview on the relationships between the classes defined by various reducibilities to sparse sets is given in Section 5.

A set is self-reducible, if its characteristic function is computable using "recursive function calls". If the arguments for the recursive calls are computable and the depth of the recursion is finite, clearly the self-reducible set is decidable. This leads to a very simple definition of polynomially bounded self-reducibility.

Definition 2 *Let $\prec$ be a partial order[4] on Σ^*. $\prec$ is called* polynomially well-founded, *if there exists a polynomial p such that for all $x, y \in \Sigma^*$, $x \prec y$ implies that $|x| \leq p(|y|)$. A set A is (polynomial time)* self-reducible under $\prec$, *if there is some polynomial-time oracle machine M, such that $A = L(M, A)$ and, on every input x, M queries the oracle only about words $q \prec x$. A is* poynomially well-founded self-reducible, *if A is self-reducible under a polynomially well-founded order $\prec$.*

---

[4] i.e. an irreflexive, transitive, and antisymmetrical relation

For a polynomially well-founded $\prec$ every "increasing chain" $x_0 \prec x_1 \prec \cdots \prec x_k$ consists of at most $2^{p(|x_k|)+1}$ elements from $\Sigma^{\leq p(|x_k|)}$ for a polynomial p. For the most "popular" types of orders for self-reducibilities – the length-decreasing and the lexicographic order (see [Ba]) as well as the polynomially related order (see [Ko1]) – this upper bound underlies a stronger restriction.

Definition 3 [Ko1] *A partial order $\prec$ on Σ^* is said to be* polynomially related, *if there is a polynomial p such that (1) $x \prec y$ is decidable in time polynomial in $|x| + |y|$, and (2) for all $x_1, x_2, \ldots, x_k \in \Sigma^*$: if $x_1 \prec x_2 \prec \cdots \prec x_k$, then $k \leq p(|x_k|)$.*

Clearly, every length-decreasing order is polynomially related, whereas this is not the case for the lexicographic order.

Definition 4 [Ko1, Ba] *A set A is* length-decreasing (resp. word-decreasing, polynomially related) *self-reducible, if A is self-reducible under a length-decreasing (resp. lexicographic, polynomially related) ordering.*

One method to decide a self-reducible set $A = L(M, A)$ is to handle all oracle queries of the self-reduction machine as "recursive function calls". This leads to the following upper bounds for the complexity of self-reducible sets.

Proposition 5
1. *Every polynomially well-founded self-reducible set is in EXPSPACE.*
2. *[BBS] Every polynomially related or length-decreasing self-reducible set is in PSPACE.*
3. *[Ba] Every word-decreasing self-reducible set is in ETIME.*[5]

One also can reverse the "top down" decision method for a self-reducible set A into a "bottom up" method, if one knows the ordering $\prec$ which restricts the queries of the self-reduction machine M. Lozano and Torán [LT] showed how this "bottom up" decision process for a polynomially related or word-decreasing self-reducible set can be shortened if the set itself is sparse or many-one reducible to a sparse set. In their proofs the complexity of the computation of the minimal strings played an important role. Therefore we now define a property of orderings dependent on that complexity.

Definition 6 *Let B and C be sets and $\prec$ be an ordering. We say that a string x is* minimal in B for C *w.r.t. $\prec$, if (1) $x \in B$, and (2) $\forall z : z \prec x \Rightarrow z \notin B$, and (3) $\exists y \in C : x \prec y$. We say that a polynomially well-founded ordering is* FP^{NP}-minimizable, *if there exists an FP^{NP} transducer M using an oracle, such that for all B and i it holds that $M^B(0^i)$ outputs an x which is minimal in B for Σ^i w.r.t. $\prec$. A set A is* FP^{NP}-minimizable *self-reducible, if A is polynomial time self-reducible under an FP^{NP}-minimizable ordering $\prec$.*

Proposition 7 *The lexicographic order and every polynomially related order are FP^{NP}-minimizable.*

---

[5] $ETIME = \bigcup_{i \geq 0} DTIME(2^{cn})$

This can be shown using a prefix search technique and Kadin's census technique [Kad]. So the new type covers all of the standard types of self-reducibility, and also the so called left-sets introduced in [OW].

Proposition 8 *If a set A is a left-set, or length-decreasing, word-decreasing, or polynomially related self-reducible, then A is FP^{NP}-minimizable self-reducible.*

3   Reduction classes to sparse sets

We will show now how APT' and P-*close* fit into the context of reduction classes to sparse sets. Let $SPARSE_P$ be the class of all subsets of sparse sets in P.

Theorem 9. $R_m^p(APT') = R_m^p(SPARSE_P)$.

Using Theorem 9 we immediately get

Theorem 10. $R_m^p(APT') \subseteq IC[\log, \text{poly}]$.

The next Theorem solves an open problem from [Ko1].

Theorem 11. $R_m^p(\text{P-close}) = R_{1-tt}^p(SPARSE)$.

4   Complexity of self-reducible sets

For self-reducible sets which reduce to sparse sets the first upper bound for their complexity follows from Karp and Lipton [KL]. Lozano and Torán [LT] considered the effect of different self-reducibilities for sets in more restricted reduction classes to sparse sets.

Theorem 12.   *1.* **[BBS, Ba]** *Every length-decreasing or word-decreasing self-reducible set in $R_{tt}^p(SPARSE)$ is low in Σ_2^p.*
   2. **[AKM2]** *Every length-decreasing self-reducible set in $R_{hd}^p(R_c^p(SPARSE))$ is low in Δ_2^p.*
   3. **[LT]** *Every word-decreasing or polynomially related self-reducible set in $R_m^p(SPARSE)$ is low in Δ_2^p.*
   4. **[KOSW]** *Every length-decreasing self-reducible set in $IC[\log, \text{poly}]$ is in P.*

We now state the new results about self-reducible sets. The following proposition is easy to prove using a "top-down" method.

Proposition 13 *Every polynomially well-founded self-reducible set in APT' is in P.*

Since every set in P is self-reducible and $P \subseteq APT'$ we get

Proposition 14 *P equals the class of all polynomially well-founded self-reducible sets in APT'.*

We now show that all polynomially well-founded self-reducible sets in IC[log,poly] are in $\mathrm{NP} \cap \mathrm{co\text{-}NP}$. Compared with the result from [KOSW] as cited in Theorem 12 the complexity of sets in IC[log, poly] which are decidable using exponential deep recursion is not much greater than that of sets in IC[log, poly] decidable by polynomially deep recursion.

Theorem 15. *Every polynomially well-founded self-reducible set in* $R_c^p(\mathrm{SPARSE}) \cap R_d^p(\mathrm{co\text{-}SPARSE})$ *is in* $\mathrm{NP} \cap \mathrm{co\text{-}NP}$.

Proof Sketch Let $A \in R_c^p(\mathrm{SPARSE}) \cap R_d^p(\mathrm{co\text{-}SPARSE})$ be a polynomially well-founded self-reducible set. Then there exist a sparse set S and FP functions f and g such that $A \leq_c^p S$ via f, and $\overline{A} \leq_c^p S$ via g. Remark that for all x, $f(x) \subseteq S \iff g(x) \not\subseteq S$. Let the polynomially well-founded self-reducibility of A be witnessed by the machine M and the ordering $\prec$. Let p be a polynomial which bounds $\prec$ and the running time of f, g, and M. Consider the following Turing machine M' which on an input consisting of a string x and a (finite) set T accepts, rejects, or halts without decision.

> **input** $\langle x, T \rangle$
> simulate M on input x, but whenever M queries q to the oracle then
> > **if** $f(q) \subseteq T \not\Leftrightarrow g(q) \not\subseteq T$ **then** *halt without decision*
> > **else if** $f(q) \subseteq T$
> > > **then** continue with answer "yes"
> > > **else** continue with answer "no" (* $g(q) \subseteq T$ *)
> > **end**
> **end** (* of the simulation in case of a query of M to the oracle *)
> **if** M accepts **then** *accept* **else** *reject* **end**

The machine runs in polynomial time, and it is "robust" in the sense that for every $T \subseteq S$ the decision of M' on input $\langle x, T \rangle$ answers $x \in A$, if it decides.

Claim 1 *Let* $T \subseteq S$. *Then for all* x, *if* $M'(\langle x, T \rangle)$ *decides its input, then* $M'(\langle x, T \rangle)$ *accepts iff* $x \in A$.

and let $Queries(x_1, \ldots, x_n) = \bigcup_{1 \leq i \leq n} query(x_i)$. By definition, $query(x) \subseteq S$ for all x. There exists a short sequence of strings whose $Queries$-set can be used as oracle to decide all instances of A up to a certain length.

Claim 2 *For all* n, *there exist* $m \leq p(p(n))$ *and a sequence* $x_1, \ldots, x_m$ *of strings in* $\Sigma^{\leq p(n)}$, *such that for all* $x \in \Sigma^{\leq n}$ *it holds that* $f(x) \subseteq Queries(x_1, \ldots, x_m) \iff g(x) \not\subseteq Queries(x_1, \ldots, x_m)$ *and* $M'(\langle x, Queries(x_1, \ldots, x_m) \rangle)$ *decides.*

The following Turing machine N decides A in the sense of strong non-determinism.

> **input** y
> $T := \emptyset$
> **guess** $m \leq p(p(|y|))$, $x_1, \ldots, x_m \in \Sigma^{\leq p(|y|)}$

for $i := 1$ **to** m **do**
  if $M'(\langle x_i, T \rangle)$ halts without decision
    then halt without decision
    else if $M'(\langle x_i, T \rangle)$ accepts **then** $T := T \cup f(x_i)$ **else** $T := T \cup g(x_i)$
  end
end (* of the for-loop *)
if $M'(\langle y, T \rangle)$ accepts **then** *accept* **else if** $M'(\langle y, T \rangle)$ rejects **then** *reject*
end

By the above Claims, N is a strong non-deterministic polynomial time bounded Turing machine deciding A. ∎

Corollary 16 *Every polynomially well-founded self-reducible set in* $R_m^{co-np}(\text{SPARSE}) \cap R_m^{np}(\text{co-SPARSE})$ *is in* P^{NP}.

Since every set in NP ∩ co-NP is low in NP [Sch1] it follows that

Theorem 17. *Every polynomially well-founded self-reducible set in* IC[log, poly] *is low in* NP.

Now we consider self-reducible sets which disjunctively reduce to sparse sets. Lozano and Torán [LT] showed that in some relativized world there exists a self-reducible sparse set which is not low for Θ_2^p. Therefore, to prove lowness for Δ_2^p of self-reducible sets which reduce to sparse sets is the "best" we can hope for.

Theorem 18. *For every* FP^{NP}-*minimizable self-reducible set A in* $R_d^p(\text{SPARSE})$ *there exists an* FP^{NP}-*printable sparse set S' such that* $A \leq_d^p S'$.

Proof Sketch Let $A \in R_d^p(\text{SPARSE})$. From an observation in [AKM1] it follows that $A \leq_d^p S$ via a function f, such that for a strictly increasing polynomial r and a polynomial t, $f(x) \subseteq \Sigma^{r(|x|)}$ and $|f(x)| \leq t(|x|)$ for all $x \in \Sigma^*$. Let A be an FP^{NP}-minimizable self-reducible set, witnessed by the machine M and the ordering $\prec$. The following polynomially time bounded machine M' which gets as input a string x and a set O and simulates M using the disjunctive query mechanism to O to answer the oracle queries.

  input $\langle x, O \rangle$
  simulate M on input x, but whenever M queries q to the oracle then
    if $f(q) \cap O \neq \emptyset$
      then continue with answer "yes"
      else continue with answer "no"
    end
  accept if and only if M accepts

We can use the inductive structure of $\prec$ to derive that M' on input $\langle x, O \rangle$ decides the membership of x w.r.t. A correctly if x and O have certain properties.

Claim 3 *For all $x \in \Sigma^*$ and $Q \subseteq \Sigma^*$: if for all y with $y \prec x$ it holds that* $\langle y, Q \rangle \in L(M') \iff f(y) \cap Q \neq \emptyset$, *then for all y with $y \prec x$ or $y = x$ it follows that* $\langle y, Q \rangle \in L(M') \iff y \in A$.

Let $err(\langle x, Q \rangle) = $ "$\langle x, Q \rangle \in L(M') \not\Leftrightarrow f(x) \cap Q \neq \emptyset$". Since err is in P, and since $\prec$ is FP^{NP}-minimizable, the function h which on input $\langle 0^i, Q \rangle$ outputs a minimal string in B for Σ^i is in FP^{NP}. The following function CONSTRUCT computes on input 0^i a sparse set S_i. The union over all those S_i is a sparse subset of $\bigcup_{x \in A} f(x) - \bigcup_{x \notin A} f(x)$.

CONSTRUCT(0^i)
    $Q := \emptyset$
    while $\exists y \in \Sigma^i\colon err(\langle y, Q \rangle)$ **do**
    $x := h(\langle 0^i, Q \rangle)$
    if $\langle x, Q \rangle \in L(M')$ **then** $Q := Q \cup f(x)$ **else** $Q := Q - f(x)$ **end**
    end (* while *)
    return Q

It follows that the set $S' = \bigcup_{i \geq 0}(\text{CONSTRUCT}(0^i) \cap \Sigma^{r(i)})$ is sparse, that $A \leq^p_d S$ via f, and that S' is FP^{NP}-printable. ∎

Finally we consider class $R^p_{hd}(R^p_c(\text{SPARSE}))$ which contains the Boolean closure of $R^p_c(\text{SPARSE})$ (see [AKM1]).

Theorem 19.
For every FP^{NP}-minimizable self-reducible set A in $R^p_{hd}(R^p_c(\text{SPARSE}))$, there exists a sparse FP^{NP}-printable set S' such that $A \in R^p_{hd}(R^p_c(S'))$.

The proof is a straightforward modification of a proof from [AKM2] where a similar result is shown for length-decreasing self-reducible sets.

Theorem 20. *Every FP^{NP}-minimizable self-reducible set in $R^p_{hd}(R^p_c(\text{SPARSE}))$, in $R^p_d(\text{SPARSE})$, or in $R^p_c(\text{co-SPARSE})$ is low in Δ^p_2.*

By Proposition 8 the same holds for word-decreasing self-reducible sets. Since the NP-complete set SAT and and the PSPACE-complete set QBF are length-decreasing self-reducible we get

Proposition 21

1. *If SAT $\in R^p_d(\text{SPARSE}) \cup R^p_c(\text{co-SPARSE}) \cup R^p_{hd}(R^p_c(\text{SPARSE}))$, then the Polynomial Time Hierarchy collapses to Δ^p_2.*
2. *If QBF $\in R^p_d(\text{SPARSE}) \cup R^p_c(\text{co-SPARSE}) \cup R^p_{hd}(R^p_c(\text{SPARSE}))$, then PSPACE $= \Delta^p_2$.*

5 Summary and overview

Figure 1 summarizes the results of this paper and shows their relations to previously known results. It gives an overview on the relationships of the reduction classes to sparse sets, the collapses of the Polynomial Time Hierarchy PH under the assumption that NP is contained in one of these classes, the extended

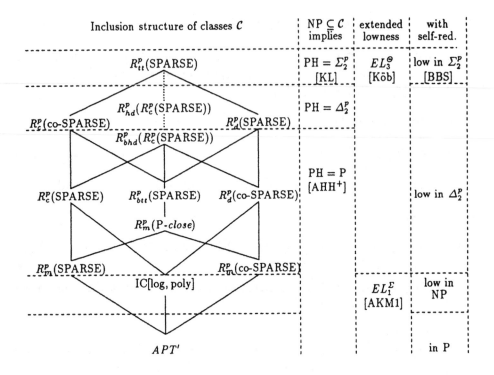

Fig. 1. Overview

lowness (as defined in [BBS]) of the reduction classes, and on the lowness of self-reducible sets in these classes. Dotted lines in the inclusion diagram stand for $\subseteq$, the other inclusions are proper ($\not\subseteq$) (cf. [Ko2, AHH$^+$, AKM1, BLS]).

Acknowledgements

I like to thank Ricard Gavaldà and Johannes Köbler for helpful comments and discussions.

References

[AHH$^+$] V. Arvind, Y. Han, L. Hemachandra, J. Köbler, A. Lozano, M. Mundhenk, M. Ogiwara, U. Schöning, R. Silvestri, and T. Thierauf. Reductions to sets of low information content. In *Complexity Theory*, K. Ambos-Spies, S. Homer, and U. Schöning (eds.). Cambridge University Press, 1993.

[AKM1] V. Arvind, J. Köbler, and M. Mundhenk. Lowness and the complexity of sparse and tally descriptions. *Proc. 3rd ISAAC*, Lecture Notes in Computer Science, #650:249–258, Springer Verlag, 1992.

[AKM2] V. Arvind, J. Köbler, and M. Mundhenk. Hausdorff reductions to sparse sets and to sets of high information content. *Proc. 18th MFCS*, Lecture Notes in Computer Science, #711:232–241, Springer Verlag, 1993.

[Ba]   J. Balcázar. Self-reducibility. *Journal of Computer and System Sciences*, 41:367-388, 1990.

[BBS]   J.L. Balcázar, R. Book, and U. Schöning. Sparse sets, lowness and highness. *SIAM Journal on Computing*, 15:739-747, 1986.

[BDG]   J.L. Balcázar, J. Díaz, and J. Gabarró. *Structural Complexity I.* EATCS Monographs on Theoretical Computer Science, Springer Verlag, 1988.

[Ber]   P. Berman. Relationship between density and deterministic complexity of NP-complete languages. *Proceedings of the 5th ICALP*, Lecture Notes in Computer Science, #62:63-71, Springer Verlag, 1978.

[BL]   R. Book and J. Lutz. On languages with very high space-bounded Kolmogorov complexity. *SIAM Journal on Computing*, 22:395-402, 1993.

[BLS]   H. Buhrman, L. Longpré, and E. Spaan. Sparse reduces conjunctively to tally. *Proceedings of the 8th Structure in Complexity Theory Conference*, IEEE Computer Society Press, 1993.

[Kad]   J. Kadin. $P^{NP[\log n]}$ and sparse Turing-complete sets for NP. *Journal of Computer and System Sciences*, 39(3):282-298, 1989.

[KL]   R. Karp and R. Lipton. Some connections between nonuniform and uniform complexity classes. *Proceedings of the 12th ACM Symposium on Theory of Computing*, 302-309, 1980.

[Ko1]   K. Ko. On self-reducibility and weak p-selectivity. *Journal of Computer and System Sciences*, 26:209-221, 1983.

[Ko2]   K. Ko. Distinguishing conjunctive and disjunctive reducibilities by sparse sets. *Information and Computation*, 81(1):62-87, 1989.

[KOSW] K. Ko, P. Orponen, U. Schöning, and O. Watanabe. Instance complexity. *Journal of the ACM*, to appear.

[Köb]   J. Köbler. Locating P/poly optimally in the extended low hierarchy. *Proc. of the 10th STACS*, Lecture Notes in Computer Science, #665:28-37, Springer Verlag, 1993.

[LT]   A. Lozano and J. Torán. Self-reducible sets of small density. *Mathematical Systems Theory*, 24:83-100, 1991.

[MP]   A. Meyer, M. Paterson. With what frequency are apparently intractable problems difficult? Tech. Report MIT/LCS/TM-126, Lab. for Computer Science, MIT, Cambridge, 1979.

[OW]   M. Ogiwara and O. Watanabe. On polynomial-time bounded truth-table reducibility of NP sets to sparse sets. *SIAM Journal on Computing*, 20(3):471-483, 1991.

[Sch1] U. Schöning. A low and a high hierarchy within NP. *Journal of Computer and System Sciences*, 27:14-28, 1983.

[Sch2] U. Schöning. *Complexity and Structure*, Lecture Notes in Computer Science, #211, Springer Verlag, 1985

[Wag]   K.W. Wagner. More complicated questions about maxima and minima, and some closures of NP. *Theoretical Computer Science*, 51:53-80, 1987.

Lower Bounds for Merging on the Hypercube

Christine Rüb

Max-Planck Institut für Informatik, Im Stadtwald, D-66123 Saarbrücken, Germany

Abstract We show lower bounds for the problems of merging two sorted lists of equal length and sorting by repeatedly merging pairs of sorted sequences on the hypercube. These lower bounds hold on the average for any ordering of the processors of the hypercube.

Key Words Hypercube, Merging, Sorting, Lower bounds.

1. Motivation and Introduction

The development of a deterministic algorithm for sorting on the hypercube that is work-optimal[1] and runs in polylogarithmic time is a long-standing open problem and has attracted considerable interest (see, e.g. [AH88], [CyP90], [CyS92], [LP90], [P89], [P92]). There are, however, several work-optimal deterministic algorithms for sorting on the PRAM-model that run in polylogarithmic time (see, e.g. [BN89], [C88], [HR89], [K83]). All of these algorithms have in common that they sort by repeatedly merging pairs of sorted sequences. Thus a question to ask is: is the same possible for the hypercube, i.e. is it possible to merge two sorted lists of altogether n elements using $O(n)$ work and polylog(n) time? In this paper we show that this is not the case.

In particular we show the following: Let $\mathcal{H}$ be an n-node hypercube [2] with processors $P_0,...,P_{n-1}$. Let A and B be two sorted sequences of length $n/2$ each such that the elements of A (B, resp.,) are stored in a sorted order at the first (last, resp.,) $n/2$ processors of $\mathcal{H}$. Assume that we want to merge A and B, i.e. we want to rearrange the elements in $A \cup B$ such that the i-th largest element in this sequence is stored at the i-th processor of $\mathcal{H}$. Then there are instances of A and B such that the number of traversed edges, summed over all elements in $A \cup B$, is at least $(n \log n)/2$. Since in each step a maximum of n edges can be used (one per processor), this means that $(\log n)/2$ steps are necessary to merge the two sequences. Thus sorting n elements by repeatedly using ("standard") two-way merging requires work $\Omega(n(\log p)^2)$ on a p-node hypercube, and cannot be optimal unless $p = O(2^{\sqrt{\log n}})$ and

---

[1] A parallel algorithm is work-optimal if the product of the number of processors used and the running time is of the same order as the running time of the best sequential algorithm for the same problem.

[2] A d-dimensional hypercube consists of $n = 2^d$ processors P_0 to P_{n-1}. Processor P_i is connected to all processors P_j where the binary representation of i and j differ in exactly one bit, i.e. where the hamming distance between i and j is one.

This work was supported by the DFG, SFB 124, TP B2, VLSI Entwurfsmethoden und Parallelität.

$t = \Omega((n\log n)/2^{\sqrt{\log n}})$. This also rules out a pipelined merge-sort as in the PRAM algorithm of [C88].

However, does this really show that sorting by (two-way) merging cannot be optimal? Above we required that the i-th element of A resides initially at the i-th processor of $\mathcal{H}$ and the i-th element of B at the $(n/2+i)$-th processor of $\mathcal{H}$, and that at the end the i-th element of $A \cup B$ resides at the i-th processor of $\mathcal{H}$, i.e. we assumed that the processors are ordered by their indices. So, what happens if we use a different ordering of the processors? Can we find an ordering of them such that two-way merging of n elements on an n-node hypercube can be done in $o(n \log n)$ work? (In this paper work means the total number of executed steps; idle time of processors is not counted.) If this were the case, we could, perhaps, use this ordering to sort the given elements in $O(n \log n) + o(n(\log n)^2)$ work and then rearrange the elements according to the standard ordering of processors in additional $O(n \log n)$ work, see, e.g. [L92], pp. 451 ff. Alas, it is of no use to change the ordering of the processors: we will show that two-way merging of n elements on an n-node hypercube needs at least work $(n \log n)/4$ in the worst case and $\Omega(n \log n)$ on the average (over all possible outcomes of the merging) regardless of the ordering of the processors, and that sorting n elements by two-way merging needs work $\Omega(n(\log p)^2)$ on a p-node hypercube if $p \geq n^{0.5+\epsilon}$ for every $\epsilon > 0$. The latter holds even if the ordering of the processors changes in each step of the recursion, as long as these orderings are independent of the input.

Note that the almost optimal algorithm in [CyP90] that sorts n elements on an n-node hypercube in time $O(\log n \log \log n)$ does this by merging $\sqrt{n}$ sorted lists of $\sqrt{n}$ elements each.

To prove the above lower bounds we make use of the fact that we want to compute the elements in $A \cup B$ in a certain order, i.e. that we want to rearrange the input elements. If we are content with computing the rank of each element in $A \cup B$, this is no longer the case. Nevertheless, we will show similar lower bounds for computing only the ranks of the elements.

We are not aware of any other work on this subject. For lower bounds on the size of comparator networks for merging we refer to [MPT92].

This paper is organized as follows. Section 2 explains the general idea, Section 3 shows that there are inputs that cause high running times, Section 4 shows that the average running time for merging is high, and Section 5 considers the problem of computing the ranks.

2. The General Idea

Consider the problem of merging two lists of $n/2$ elements each on an n-node hypercube such that each processor holds one element. The ordering of the processors is arbitrary. It is easy to see that there are two input sequences such that for at least one element in them the hamming distance between the processors that hold it before and after the merging, say processors P_i and P_j, is $\Omega(\log n)$. Thus this element has to travel across $\Omega(\log n)$ edges and it will take $\Omega(\log n)$ time till the last element has reached its destination.

Next consider the case where each processor holds $m > 1$ elements and the elements at each processor form a consecutive subsequence. We can construct two input sequences such that all elements stored at P_i have to travel to P_j. But that does not necessarily mean that merging these two sequences needs time $\Omega(m \log n)$; rather, by pipelining the movement of the elements at P_i, it could be possible to achieve a running time of $O(m + \log n)$. Thus the above argument is too weak.

Instead we will use arguments of the following kind: Let $\mathcal{H}$ be an n-node hypercube. Consider the problem of merging two sorted sequences of $n/2$ elements each, one per processor, using any ordering of the processors. We will show that for a constant fraction of the input elements the average distance (over all possible outcomes of the merging) this element has to travel is $\Omega(\log n)$. Thus the average number of edges crossed by all elements together is $\Omega(n \log n)$ and for a constant fraction of all possible inputs the total number of crossed edges is $\Omega(n \log n)$. Since in each step at most n edges can be used, the running time for these inputs is $\Omega(\log n)$. Using a similar argument we will show that that merging mn, $m \geq 1$, elements on an n-node hypercube needs time $\Omega(m \log n)$ and that pipelining cannot improve this. Since this lower bound on the average number of crossed edges still holds if n', $n' \leq n$, elements reside at n' nodes of an n-node hypercube, $n = O((n')^c)$ for a constant c, this means that sorting n elements on a p-node hypercube by repeated two-way merging needs time $\Omega((n/p)(\log p)^2)$ and work $\Omega(n(\log p)^2)$ if $p \geq n^{0.5+\epsilon}$ for any constant $\epsilon > 0$.

3. Expensive Inputs

In this section we consider the problem of merging n elements on an n-node hypercube and show that there are inputs that cause a high running time. Namely we show that for every ordering of the processors there is an input with a running time of at least $(\log n + \sqrt{\log n/2})/4$, and for the standard ordering we give an input with a running time of at least $(\log n)/2$.

First we consider arbitrary orderings. Let $A[0..n/2-1]$ and $B[0..n/2-1]$ be the two sequences to be merged, and let $C[0..n-1]$ be the output of the merging. At the beginning (the end, resp.) each of the n processors holds exactly one element of $A \cup B$ (of C, resp.). We consider the following $n/2$ inputs $I_0, ..., I_{n/2-1}$: In I_i the element $A[j]$ is moved to $C[j+i]$, $0 \leq j \leq n/2 - 1$. We will show that for every possible ordering of the processors one of these inputs needs at least $(\log n + \sqrt{\log n/2})/4$ time.

Let $d = \log n$. Consider a fixed input element $A[j]$ and let $A[j]$ reside at P^j at the beginning. In $I_0, ..., I_{n/2_1}$ there are $n/2$ different positions for $A[j]$ in C, i.e. $n/2$ different processors $A[j]$ has to be moved to. Assume that these processors are chosen such that the costs for $A[j]$ are minimized, i.e they are the processors with the $n/2$ smallest hamming distances to P^j. Thus the total costs for $A[j]$ in the $n/2$ inputs are

$$\geq \sum_{i=0}^{\lfloor d/2 \rfloor} i \binom{d}{i} \qquad \text{if } d \text{ is odd, and}$$

$$\geq \sum_{i=0}^{d/2-1} i\binom{d}{i} + \frac{1}{4}d\binom{d}{d/2} \quad \text{if } d \text{ is even.}$$

Further, if d is odd,

$$\sum_{i=0}^{\lfloor d/2 \rfloor} i\binom{d}{i} = \sum_{i=1}^{(d-1)/2} d\binom{d-1}{i-1} = \frac{1}{4}d2^d + \frac{1}{2}d\binom{d-1}{(d-1)/2} \geq \frac{1}{4}(d + \sqrt{d/2})2^d,$$

and if d is even,

$$\sum_{i=0}^{d/2-1} i\binom{d}{i} + \frac{1}{4}d\binom{d}{d/2} = \frac{1}{4}d2^d + \frac{1}{4}d\binom{d}{d/2} \geq \frac{1}{4}(d + \sqrt{d/2})2^d.$$

(For the inequalities we used Sterling's approximation for $n!$.)

Thus, the total costs for all $n/2$ inputs, summed over all $A[j]$, $0 \leq j \leq n/2-1$, are at least $(n/2)(d + \sqrt{d/2})n/4$, and for at least one of the $n/2$ inputs the total costs (or work) are at least $(d + \sqrt{d/2})n/4$. Thus the running time for this input is at least $(d + \sqrt{d/2})/4 = (\log n + \sqrt{\log n/2})/4$. Note that this lower bound also holds if the orderings of the processors used at the beginning and at the end may differ.

If we use the standard ordering of the processors we can improve upon the constants in the above lower bound. Let A, B, and C be defined as above. At the beginning $A[i]$ resides at processor P_i and $B[i]$ at processor $P_{n/2+i}$, $0 \leq i \leq n/2 - 1$, and at the end $C[i]$ resides at processor P_i, $0 \leq i \leq n - 1$. If the input elements move as follows, at least $(n \log n)/2$ edges have to be crossed altogether:

$$A[i] \to C[2i], 0 \leq i \leq n/2 - 1, \text{ and}$$
$$B[i] \to C[2i + 1], 0 \leq i \leq n/2 - 1.$$

By induction on $\log n$ we can show that the performed work is at least $(n \log n)/2$. Applying this lower bound on each level of a two-way merge sort shows that such a sort algorithm performs a work of at least $n \log n(\log n+1)/4$.

4. Merging is Expensive on the Average

In this section we will show that merging two sorted sequences (of equal length), using an arbitrary ordering of the processors, is expensive on the average. (We average over all possible outcomes of the merging). This is the case even if the orderings of the processors used before and after the merging may differ.

First we consider the case where there is at most one element per processor; later we will extend this to the case where the number of processors is smaller than the number of elements. Thus, let $\mathcal{H}$ be a p-node hypercube, and let A and B be two sorted sequences of size $n/2$ each, $p = n^c$ for a constant $c \geq 1$, and n a power of 2. Assume that A and B are stored at the nodes of a subset S_A (S_B, resp.,) of the processors of $\mathcal{H}$, $|S_A| = |S_B| = n/2$ (one element per processor), and that we want to merge A and B such that the elements in $A \cup B$ are afterwards stored at a subset S of the processors, $|S| = n$, again one element per processor. The elements are stored in S_A, S_B, and S according to some fixed ordering of the processors in these sets. Note that S_A, S_B, and S need not be disjoint. (This corresponds to a merge step in a 2-way merge sort.)

We start off from the following idea. Let $M(n)$ be the number of possible outcomes of merging two lists of size $n/2$ each. For each input element each of these outcomes has some costs assigned to it, namely the hamming distance between the two processors storing the element before and after the merging. Consider a fixed input element X, stored at some processor P. We want to determine the set S of processors and the ordering of these processors such that the sum of the costs over all possible outcomes of the merging is minimized for X. Let S_X denote such a choice of S and let $R_X : S_X \to \{1, ..., n\}$, R_X bijective, be the ordering of the processors in S_i. We will show that, for a constant fraction of the input elements, the sum of the costs is $\Omega(M(n) \log n)$ when S_X and R_X are used. Thus for these input elements the sum of the costs is always $\Omega(M(n) \log n)$, regardless of the choice of S and the ordering of the processors in S. Since this is true for a constant fraction of the input elements, the sum of the overall (for all input elements) costs is $\Omega(M(n) n \log n)$ and thus the average costs of merging are $\Omega(n \log n)$.

The following bounds on binomial coefficients and sums of binomial coefficients will be needed later on.

Lemma 1

Let $n \in \mathbb{N}$ and let $\mu n \in \mathbb{N}$, $0 < \mu < 1$. Then

$$\frac{1}{\sqrt{8n\mu(1-\mu)}} 2^{nH_2(\mu)} \leq \binom{n}{\mu n} \leq \frac{1}{\sqrt{2\pi n\mu(1-\mu)}} 2^{nH_2(\mu)}, \text{ and}$$

$$\frac{1}{\sqrt{8n\mu(1-\mu)}} 2^{nH_2(\mu)} \leq \sum_{k=0}^{\mu n} \binom{n}{k} \leq 2^{nH_2(\mu)}, \text{ if } 0 < \mu < 1/2,$$

where $H_2(x) = -x \log x - (1-x) \log(1-x)$.

Proof: See, e.g. [WS78], pp. 308 ff.   ∎

Let us next evaluate the number $M(n)$ of possible outcomes of merging two lists of size $n/2$ each.

Lemma 2

$$M(n) = \binom{n}{n/2}, \text{ and } \sqrt{\frac{1}{2n}} 2^n \leq \binom{n}{n/2} \leq \sqrt{\frac{2}{\pi n}} 2^n$$

Proof: This follows from Lemma 1.   ∎

Let $A = A[0..n/2 - 1]$ and $B = B[0..n/2 - 1]$ and let $C[0..n-1]$ be the result of the merging. We want to derive a lower bound on the average costs caused by the elements in A and B. To do this, we concentrate on the elements in A, since the situation for A and B is symmetrical. For each element $A[i]$, $0 \leq i \leq n/2 - 1$, there are $n/2$ possibilities for its location in C: it can move between 0 (when no element in B is smaller than $A[i]$) and $n/2$ (when all elements in B are smaller than $A[i]$) positions to the right. Each of these moves has some costs associated with it: $costs(i, b)$, $0 \leq b \leq n/2$, is the hamming distance between the processor that stores $A[i]$ and the processor that stores $C[i + b]$. Each of these costs arises in several outcomes of the merging: $costs(i, b)$ arises in all outcomes where $A[i]$ moves b positions to the right. Let the number of these outcomes be $Z(i, b)$. Then the overall costs, in all possible outcomes of the merging, caused by $A[i]$ are

$$\sum_{0 \leq b \leq n/2} costs(i, b)Z(i, b).$$

($costs(i, b)$ of course depends on the chosen orderings on the processors, whereas $Z(i, b)$ is independent of them.)
We want to derive a lower bound on

$$\min_{S, \Pi_S} \left\{ \sum_{0 \leq b \leq n/2} costs(i, b)Z(i, b) \right\} =: Min(i),$$

where Π_S denotes the ordering of the processors in S. Note that this value is independent of the index of the processor that holds $A[i]$. To be able to derive a lower bound, we next examine $Z(i, b)$.

Lemma 3

$$(1) \qquad Z(i, b) = \binom{i + b}{i} \binom{n - (i + b) - 1}{n/2 - i - 1}$$

$$(2) \qquad Z(i, b) = Z(n/2 - i - 1, n/2 - b), \quad 0 \leq i \leq n/2, \ 0 \leq b \leq n/2$$

(3) For a fixed i, $Z(i, b)$ first increases monotonically and then decreases monotonically, with a maximum at $Z(i, i)$ if $i \leq n/4 - 1$ and at $Z(i, i+1)$ else.

Proof: We omit the proof. ∎

Because of Lemma 3.2 we can concentrate on the i where $0 \leq i \leq n/4 - 1$. Next we will estimate how large a fraction of all possible outcomes the largest "weight", $Z(i, i)$, comprises.

Lemma 4

$$\frac{\pi}{2}\sqrt{i} \leq \frac{M(n)}{Z(i, i)} \leq \frac{16}{\sqrt{\pi}}\sqrt{i}, \text{ if } n \geq 4.$$

Proof: This follows from Lemmas 1 through 3. ∎

That means that the largest existing weight is proportional to $M(n)/\sqrt{i}$. To achieve $Min(i)$ for a fixed i we should choose S and the ordering of the processors in S such that the largest weights are assigned to the smallest hamming distances, e.g. $A[i]$ and $C[2i]$ should be stored at the same processor. Assume that we assign the $L = \pi\sqrt{i}/(2c)$ largest weights ($c > 1$ a constant) to the L cheapest movements. The size of each of these weights is at most $2M(n)/(\pi\sqrt{i})$, and the sum of their sizes is at most $M(n)/c$. Thus the sum of the sizes of the remaining weights, i.e. the number of possible movements that have not yet been assigned costs, is at least $(1 - 1/c)M(n)$. These remaining weights are all assigned to costs of at least $m(i, c)$, where $m(i, c)$ are the costs of the $L + 1$ cheapest movement. Thus the sum of all costs is at least $m(i, c)(1 - 1/c)M(n)$, and the average costs are at least $m(i, c)(1 - 1/c)$. Thus, if $m(i, c) = \Omega(\log n)$, the average costs are $\Omega(\log n)$. The following lemma shows that this is indeed the case if $i = \Omega(n)$.

Lemma 5
Let $i = \Omega(n)$ and let $p = n^\alpha$ for a constant $\alpha > 1$. Then for every constant $c > 1$ there is a constant μ such that the number of movements with costs less than $\mu \log p$ is at most $\pi\sqrt{i}/(2c)$, if n is sufficiently large.

Proof: We have to show that

$$\sum_{j=0}^{\mu \log p} \binom{\log p}{j} \leq \frac{\pi\sqrt{i}}{2c} := x.$$

Since

$$\sum_{j=0}^{\mu \log p} \binom{\log p}{j} \leq 2^{\log p \, H_2(\mu)},$$

it is sufficient that

$$p^{H_2(\mu)} \leq x \text{ or}$$

$$H_2(\mu) \leq \log x / \log p.$$

However,

$$\log x / \log p = \log \frac{\pi}{2c} / \log p + (1/2) \log i / \log p \geq c'$$

for a constant $c' > 0$ if n is sufficiently large. Thus the constant μ exists. ∎
From Lemma 5 it follows that, for each constant c, $m(i, c) = \Omega(\log n)$, if $i = \Omega(n)$.

Thus we have proven the following theorem.

Theorem 1
Let $\mathcal{H}$ be a p-node hypercube, let A and B be two sorted sequences of size $n/2$ each, $p = n^c$ for a constant $c > 1$, and n a power of 2. Let A and B be stored at the nodes of a subset S_A (S_B, resp.,) of the processors of $\mathcal{H}$, $|S_A| = |S_B| = n/2$ (one element per processor), and let the result of the merging be stored at a subset S of the processors, $|S| = n$, again one element per processor. The elements are stored according to some fixed ordering of the processors used.

Then the average costs for merging A and B are $\Omega(n \log n)$.

We are also ready to prove a lower bound for sorting n elements by two-way merging on an n-processor hypercube.

Theorem 2

Sorting n elements on an n-node hypercube by repeated two-way merging needs time $\Omega((\log n)^2)$ on the average. This is true for all possible orderings of the processors in the levels of the recursion and even if the ordering may change from one level to the next as long as these orderings are independent of the input.

Proof: This follows directly from Theorem 1 by considering the upper, say, $1/2 \log n$ levels of the recursion. ∎

Next we want to examine the case where each processor may hold more than one element. First we consider the situation where we want to merge mn elements on a p-node hypercube, $p = n^\alpha$ for a constant $\alpha \geq 1$, and where, before and after the merging, each of n processors of the hypercube holds m elements. We will show that the average costs for doing this are $\Omega(mn \log p)$ if $m \leq n^\beta$ for a constant $\beta < 1$.

To do this, we reconsider the proof of Theorem 1. Lemmas 2, 3, and 4 hold if we replace n by $n' = mn$. Again we assign the cheapest movements to the largest weights. Instead of Lemma 5 we now have to show:

Lemma 6

Let $i = \Omega(mn)$, $m \leq n^\beta$ for a constant $\beta < 1$, and $p = n^\alpha$ for a constant $\alpha > 1$.

Then for every constant $c > 1$, there exists a constant μ such that the number of movements with costs less than $\mu \log p$ is smaller than $\pi \sqrt{i}/(2c)$, if n is sufficiently large.

Proof: Let $i \geq c'mn$. Since each movement is now assigned to m weights, we have to show that

$$\sum_{j=0}^{\mu \log p} m \binom{\log p}{j} \leq \frac{\pi \sqrt{i}}{2c} := x.$$

It is sufficient that

$$p^{H_2(\mu)} \leq x/m \text{ or}$$

$$H_2(\mu) \leq \log(x/m)/\log p.$$

However,

$$\log(x/m)/\log p \geq \log \frac{\pi}{2c} / \log p + 1/2 \log(c'n/m)/\log p \geq c''$$

for a constant $c'' > 0$ since $m \leq n^\beta$ and if n sufficiently large. Thus the constant μ exists. ∎

Again the lower bound for merging can be used to prove a corresponding lower bound for sorting.

Theorem 3

Sorting n elements on a p-node hypercube using repeated two-way merging needs work $\Omega(n(\log p)^2)$ if $p > n^{0.5+\epsilon}$ for any constant $\epsilon > 0$.

Proof: Consider once again the upper $(1/2)\log n$ levels of the recursion. To apply Lemma 6 on these levels the condition $n/p < p^\gamma$ for a constant $\gamma < 1$ must hold. This is equivalent to $p > n^{0.5+\epsilon}$ for a constant $\epsilon > 0$. ∎

5. Computing the Rank

In Section 4 we showed that merging two sorted sequences A and B is, under certain conditions, expensive on the average. To prove these lower bounds we made use of the fact that we wanted to rearrange the input elements according to their rank in A merged with B. Thus, if we only want to compute the rank of each input element in $A \cup B$, we cannot apply these lower bounds immediately. In this section we sketch how to change the claims and the proofs to obtain similar lower bounds if we only want to compute the ranks.

We make use of the following observation: Let $A = a[0..n/2 - 1]$ and $B = B[0..n/2 - 1]$ be the two input sequences and let $C = C[0..n - 1]$ be the result of merging A with B. Assume that an $A[i]$ has rank j in C, i.e. $A[i]$ is equal to $C[j]$, and that $C[j - 1] = B[j - i - 1]$, $0 \le i \le n/2 - 1$, $0 \le j \le n - 1$. Let $k = j - i - 1$. Then $A[i]$ and $B[k]$ have to be compared to determine the rank of $A[i]$ in C.

We use the notation of Section 4. Let $A[i]$ be stored at processor P^i and $B[k]$ at processor P^k. To compare $A[i]$ and $B[k]$ these two input elements have to meet at some processor P^l. Further, $h(P^i, P^l) + h(P^k, p^l) \ge h(P^i, P^k)$, where the function h denotes the hamming distance between the indices of the two processors. Thus we can claim costs of $h(P^i, P^k)$ for each assignment of the input elements where $B[k]$ is directly before $A[i]$ in C. We will charge these costs on $A[i]$, i.e. we will consider only the elements in A.

Following this observation we define $\overline{Z}(i, b)$ as the number of all inputs where $A[i]$ has rank $i + b$ in C and $B[b - 1]$ has rank $i + b - 1$ in C, $0 \le i \le n/2 - 1$, $1 \le b \le n/2$. As in Section 4, we want to derive a lower bound on

$$\min_{S,\Pi_S}\left\{ \sum_{0 \le b \le n/2} costs(i, b)\overline{Z}(i, b) \right\} =: \overline{Min}(i),$$

where $costs(i, b)$ is the hamming distance between the indices of the processors that hold $A[i]$ ($B[b]$, resp.) in the beginning. Further,

$$\overline{Z}(i, b) = \binom{i + b - 1}{i}\binom{n - (i + b) - 1}{n/2 - i - 1} = \frac{b}{i + b}Z(i, b).$$

Thus,

$$Z(i, b)/3 \le \overline{Z}(i, b) \le Z(i) \text{ if } b \ge i/2, \text{ and}$$

$$M(n) \ge \sum_{b=0}^{n/2}\overline{Z}(i, b) \ge \sum_{b=i/2}^{n/2} Z(i, b)/3 \ge \frac{1}{6}\sum_{i=0}^{n/2} Z(i, b) = M(n)/6.$$

From this we can conclude that the average costs are only decreased by a constant factor if we merely want to compute the ranks of the input elements. We omit the details.

Acknowledgments

I would like to thank Torben Hagerup for helpful discussions.

References

[AH88]   A. Aggarwal, M.-D. A. Huang, *Network complexity of sorting and graph problems and simulating CRCW PRAMs by interconnection networks*, AWOC, LNCS 319, pp. 339-350, 1988

[BN89]   G. Bilardi, A. Nicolau, *Adaptive bitonic sorting: An optimal parallel algorithm for shared-memory machines*, SIAM J. Comput. 18, pp. 216–228, 1989.

[C88]   R. Cole, *Parallel merge sort*, SIAM J. Comput. 17, 770–785, 1988.

[CyP90]   R. Cypher, G. Plaxton, *Deterministic sorting in nearly logarithmic time on the hypercube and related computers*, 20th STOC, pp. 193–203, 1990

[CyS92]   R. Cypher, J.L.C. Sanz, *Cubesort: A parallel algorithm for sorting N data items with S-sorters*, J. of Algorithms 13, pp. 211-234, 1992

[HR89]   T. Hagerup, Ch. Rüb, *Optimal merging and sorting on the EREW PRAM*, Information Processing Letters 33 , pp. 181–185, 1989.

[K83]   C.P. Kruskal, *Searching, merging and sorting in parallel computation*, IEEE Transactions on Computers, Vol. C-32, 942-946, 1983.

[L92]   F.T. Leighton, *Introduction to parallel algorithms and architectures: Arrays, trees, hypercubes*, Morgan Kaufmann Publishers, San Mate, California, 1992

[LP90]   T. Leighton, C.G. Plaxton, *A (fairly) simple circuit that (usually) sorts*, 31st FOCS, Vol. I, pp. 264-274, 1990

[WS78]   F.J. MacWilliams, N.J. Sloane, *The Theory of Error Correcting Codes*, North-Holland Mathematical Library,Vol. 16, North-Holland, Amsterdam - New York - Oxford, 1978

[MPT92]   P.B. Miltersen, M. Paterson, J. Tarui, *The asymptotic complexity of merging networks*, 33rd FOCS, pp. 236-246, 1992

[P89]   C.G. Plaxton, *Load balancing, selection and sorting on the hypercube*, SPAA, pp. 64-73, 1989

[P92]   C.G. Plaxton, *A hypercubic sorting network with nearly logarithmic depth*, 24th STOC, pp. 405-416, 1992

Lecture Notes in Computer Science

For information about Vols. 1–699
please contact your bookseller or Springer-Verlag

Vol. 700: A. Lingas, R. Karlsson, S. Carlsson (Eds.), Automata, Languages and Programming. Proceedings, 1993. XII, 697 pages. 1993.

Vol. 701: P. Atzeni (Ed.), LOGIDATA+: Deductive Databases with Complex Objects. VIII, 273 pages. 1993.

Vol. 702: E. Börger, G. Jäger, H. Kleine Büning, S. Martini, M. M. Richter (Eds.), Computer Science Logic. Proceedings, 1992. VIII, 439 pages. 1993.

Vol. 703: M. de Berg, Ray Shooting, Depth Orders and Hidden Surface Removal. X, 201 pages. 1993.

Vol. 704: F. N. Paulisch, The Design of an Extendible Graph Editor. XV, 184 pages. 1993.

Vol. 705: H. Grünbacher, R. W. Hartenstein (Eds.), Field-Programmable Gate Arrays. Proceedings, 1992. VIII, 218 pages. 1993.

Vol. 706: H. D. Rombach, V. R. Basili, R. W. Selby (Eds.), Experimental Software Engineering Issues. Proceedings, 1992. XVIII, 261 pages. 1993.

Vol. 707: O. M. Nierstrasz (Ed.), ECOOP '93 – Object-Oriented Programming. Proceedings, 1993. XI, 531 pages. 1993.

Vol. 708: C. Laugier (Ed.), Geometric Reasoning for Perception and Action. Proceedings, 1991. VIII, 281 pages. 1993.

Vol. 709: F. Dehne, J.-R. Sack, N. Santoro, S. Whitesides (Eds.), Algorithms and Data Structures. Proceedings, 1993. XII, 634 pages. 1993.

Vol. 710: Z. Ésik (Ed.), Fundamentals of Computation Theory. Proceedings, 1993. IX, 471 pages. 1993.

Vol. 711: A. M. Borzyszkowski, S. Sokołowski (Eds.), Mathematical Foundations of Computer Science 1993. Proceedings, 1993. XIII, 782 pages. 1993.

Vol. 712: P. V. Rangan (Ed.), Network and Operating System Support for Digital Audio and Video. Proceedings, 1992. X, 416 pages. 1993.

Vol. 713: G. Gottlob, A. Leitsch, D. Mundici (Eds.), Computational Logic and Proof Theory. Proceedings, 1993. XI, 348 pages. 1993.

Vol. 714: M. Bruynooghe, J. Penjam (Eds.), Programming Language Implementation and Logic Programming. Proceedings, 1993. XI, 421 pages. 1993.

Vol. 715: E. Best (Ed.), CONCUR'93. Proceedings, 1993. IX, 541 pages. 1993.

Vol. 716: A. U. Frank, I. Campari (Eds.), Spatial Information Theory. Proceedings, 1993. XI, 478 pages. 1993.

Vol. 717: I. Sommerville, M. Paul (Eds.), Software Engineering – ESEC '93. Proceedings, 1993. XII, 516 pages. 1993.

Vol. 718: J. Seberry, Y. Zheng (Eds.), Advances in Cryptology – AUSCRYPT '92. Proceedings, 1992. XIII, 543 pages. 1993.

Vol. 719: D. Chetverikov, W.G. Kropatsch (Eds.), Computer Analysis of Images and Patterns. Proceedings, 1993. XVI, 857 pages. 1993.

Vol. 720: V.Mařík, J. Lažanský, R.R. Wagner (Eds.), Database and Expert Systems Applications. Proceedings, 1993. XV, 768 pages. 1993.

Vol. 721: J. Fitch (Ed.), Design and Implementation of Symbolic Computation Systems. Proceedings, 1992. VIII, 215 pages. 1993.

Vol. 722: A. Miola (Ed.), Design and Implementation of Symbolic Computation Systems. Proceedings, 1993. XII, 384 pages. 1993.

Vol. 723: N. Aussenac, G. Boy, B. Gaines, M. Linster, J.-G. Ganascia, Y. Kodratoff (Eds.), Knowledge Acquisition for Knowledge-Based Systems. Proceedings, 1993. XIII, 446 pages. 1993. (Subseries LNAI).

Vol. 724: P. Cousot, M. Falaschi, G. Filè, A. Rauzy (Eds.), Static Analysis. Proceedings, 1993. IX, 283 pages. 1993.

Vol. 725: A. Schiper (Ed.), Distributed Algorithms. Proceedings, 1993. VIII, 325 pages. 1993.

Vol. 726: T. Lengauer (Ed.), Algorithms – ESA '93. Proceedings, 1993. IX, 419 pages. 1993

Vol. 727: M. Filgueiras, L. Damas (Eds.), Progress in Artificial Intelligence. Proceedings, 1993. X, 362 pages. 1993. (Subseries LNAI).

Vol. 728: P. Torasso (Ed.), Advances in Artificial Intelligence. Proceedings, 1993. XI, 336 pages. 1993. (Subseries LNAI).

Vol. 729: L. Donatiello, R. Nelson (Eds.), Performance Evaluation of Computer and Communication Systems. Proceedings, 1993. VIII, 675 pages. 1993.

Vol. 730: D. B. Lomet (Ed.), Foundations of Data Organization and Algorithms. Proceedings, 1993. XII, 412 pages. 1993.

Vol. 731: A. Schill (Ed.), DCE – The OSF Distributed Computing Environment. Proceedings, 1993. VIII, 285 pages. 1993.

Vol. 732: A. Bode, M. Dal Cin (Eds.), Parallel Computer Architectures. IX, 311 pages. 1993.

Vol. 733: Th. Grechenig, M. Tscheligi (Eds.), Human Computer Interaction. Proceedings, 1993. XIV, 450 pages. 1993.

Vol. 734: J. Volkert (Ed.), Parallel Computation. Proceedings, 1993. VIII, 248 pages. 1993.

Vol. 735: D. Bjørner, M. Broy, I. V. Pottosin (Eds.), Formal Methods in Programming and Their Applications. Proceedings, 1993. IX, 434 pages. 1993.

Vol. 736: R. L. Grossman, A. Nerode, A. P. Ravn, H. Rischel (Eds.), Hybrid Systems. VIII, 474 pages. 1993.

Vol. 737: J. Calmet, J. A. Campbell (Eds.), Artificial Intelligence and Symbolic Mathematical Computing. Proceedings, 1992. VIII, 305 pages. 1993.

Vol. 738: M. Weber, M. Simons, Ch. Lafontaine, The Generic Development Language Deva. XI, 246 pages. 1993.

Vol. 739: H. Imai, R. L. Rivest, T. Matsumoto (Eds.), Advances in Cryptology – ASIACRYPT '91. X, 499 pages. 1993.

Vol. 740: E. F. Brickell (Ed.), Advances in Cryptology – CRYPTO '92. Proceedings, 1992. X, 593 pages. 1993.

Vol. 741: B. Preneel, R. Govaerts, J. Vandewalle (Eds.), Computer Security and Industrial Cryptography. Proceedings, 1991. VIII, 275 pages. 1993.

Vol. 742: S. Nishio, A. Yonezawa (Eds.), Object Technologies for Advanced Software. Proceedings, 1993. X, 543 pages. 1993.

Vol. 743: S. Doshita, K. Furukawa, K. P. Jantke, T. Nishida (Eds.), Algorithmic Learning Theory. Proceedings, 1992. X, 260 pages. 1993. (Subseries LNAI)

Vol. 744: K. P. Jantke, T. Yokomori, S. Kobayashi, E. Tomita (Eds.), Algorithmic Learning Theory. Proceedings, 1993. XI, 423 pages. 1993. (Subseries LNAI)

Vol. 745: V. Roberto (Ed.), Intelligent Perceptual Systems. VIII, 378 pages. 1993. (Subseries LNAI)

Vol. 746: A. S. Tanguiane, Artificial Perception and Music Recognition. XV, 210 pages. 1993. (Subseries LNAI).

Vol. 747: M. Clarke, R. Kruse, S. Moral (Eds.), Symbolic and Quantitative Approaches to Reasoning and Uncertainty. Proceedings, 1993. X, 390 pages. 1993.

Vol. 748: R. H. Halstead Jr., T. Ito (Eds.), Parallel Symbolic Computing: Languages, Systems, and Applications. Proceedings, 1992. X, 419 pages. 1993.

Vol. 749: P. A. Fritzson (Ed.), Automated and Algorithmic Debugging. Proceedings, 1993. VIII, 369 pages. 1993.

Vol. 750: J. L. Díaz-Herrera (Ed.), Software Engineering Education. Proceedings, 1994. XII, 601 pages. 1994.

Vol. 751: B. Jähne, Spatio-Temporal Image Processing. XII, 208 pages. 1993.

Vol. 752: T. W. Finin, C. K. Nicholas, Y. Yesha (Eds.), Information and Knowledge Management. Proceedings, 1992. VII, 142 pages. 1993.

Vol. 753: L. J. Bass, J. Gornostaev, C. Unger (Eds.), Human-Computer Interaction. Proceedings, 1993. X, 388 pages. 1993.

Vol. 754: H. D. Pfeiffer, T. E. Nagle (Eds.), Conceptual Structures: Theory and Implementation. Proceedings, 1992. IX, 327 pages. 1993. (Subseries LNAI).

Vol. 755: B. Möller, H. Partsch, S. Schuman (Eds.), Formal Program Development. Proceedings. VII, 371 pages. 1993.

Vol. 756: J. Pieprzyk, B. Sadeghiyan, Design of Hashing Algorithms. XV, 194 pages. 1993.

Vol. 757: U. Banerjee, D. Gelernter, A. Nicolau, D. Padua (Eds.), Languages and Compilers for Parallel Computing. Proceedings, 1992. X, 576 pages. 1993.

Vol. 758: M. Teillaud, Towards Dynamic Randomized Algorithms in Computational Geometry. IX, 157 pages. 1993.

Vol. 759: N. R. Adam, B. K. Bhargava (Eds.), Advanced Database Systems. XV, 451 pages. 1993.

Vol. 760: S. Ceri, K. Tanaka, S. Tsur (Eds.), Deductive and Object-Oriented Databases. Proceedings, 1993. XII, 488 pages. 1993.

Vol. 761: R. K. Shyamasundar (Ed.), Foundations of Software Technology and Theoretical Computer Science. Proceedings, 1993. XIV, 456 pages. 1993.

Vol. 762: K. W. Ng, P. Raghavan, N. V. Balasubramanian, F. Y. L. Chin (Eds.), Algorithms and Computation. Proceedings, 1993. XIII, 542 pages. 1993.

Vol. 763: F. Pichler, R. Moreno Díaz (Eds.), Computer Aided Systems Theory – EUROCAST '93. Proceedings, 1993. IX, 451 pages. 1994.

Vol. 764: G. Wagner, Vivid Logic. XII, 148 pages. 1994. (Subseries LNAI).

Vol. 765: T. Helleseth (Ed.), Advances in Cryptology – EUROCRYPT '93. Proceedings, 1993. X, 467 pages. 1994.

Vol. 766: P. R. Van Loocke, The Dynamics of Concepts. XI, 340 pages. 1994. (Subseries LNAI).

Vol. 767: M. Gogolla, An Extended Entity-Relationship Model. X, 136 pages. 1994.

Vol. 768: U. Banerjee, D. Gelernter, A. Nicolau, D. Padua (Eds.), Languages and Compilers for Parallel Computing. Proceedings, 1993. XI, 655 pages. 1994.

Vol. 769: J. L. Nazareth, The Newton-Cauchy Framework. XII, 101 pages. 1994.

Vol. 770: P. Haddawy (Representing Plans Under Uncertainty. X, 129 pages. 1994. (Subseries LNAI).

Vol. 771: G. Tomas, C. W. Ueberhuber, Visualization of Scientific Parallel Programs. XI, 310 pages. 1994.

Vol. 772: B. C. Warboys (Ed.),Software Process Technology. Proceedings, 1994. IX, 275 pages. 1994.

Vol. 773: D. R. Stinson (Ed.), Advances in Cryptology – CRYPTO '93. Proceedings, 1993. X, 492 pages. 1994.

Vol. 774: M. Banâtre, P. A. Lee (Eds.), Hardware and Software Architectures for Fault Tolerance. XIII, 311 pages. 1994.

Vol. 775: P. Enjalbert, E. W. Mayr, K. W. Wagner (Eds.), STACS 94. Proceedings, 1994. XIV, 782 pages. 1994.

Vol. 778: M. Bonuccelli, P. Crescenzi, R. Petreschi (Eds.), Algorithms and Complexity. Proceedings, 1994. VIII, 222 pages. 1994.